Just Transitions

Just Transitions

Social Justice in the Shift Towards a Low-Carbon World

Edited by
Edouard Morena, Dunja Krause
and Dimitris Stevis

First published 2020 by Pluto Press
345 Archway Road, London N6 5AA

www.plutobooks.com

British Library Cataloguing in Publication Data
A catalogue record for this book is available from the British Library

ISBN   978 0 7453 3994 8   Hardback
ISBN   978 0 7453 3992 4   Paperback
ISBN   978 1 7868 0532 4   PDF eBook
ISBN   978 1 7868 0534 8   Kindle eBook
ISBN   978 1 7868 0533 1   EPUB eBook

This book is printed on paper suitable for recycling and made from fully managed and
sustained forest sources. Logging, pulping and manufacturing processes are expected to
conform to the environmental standards of the country of origin.

Typeset by Stanford DTP Services, Northampton, England

Simultaneously printed in the United Kingdom and United States of America

Contents

List of illustrations

List of acronyms

AfD	Alternative für Deutschland (*Alternative for Germany, far-right populist party*)
AFL	Alberta Federation of Labour
AFL-CIO	American Federation of Labor and Congress of Industrial Organizations (United States)
AIPP	Asia Indigenous Peoples Pact
AG	Auditor General (South Africa)
ALP	Australian Labor Party
AMMA	Australian Mines and Metals Association
AMWU	Australian Manufacturing Workers' Union
ANC	African National Congress (South Africa)
APWLD	Asia Pacific Forum on Women, Law and Development
AWG-LCA	Ad Hoc Working Group on Long-term Cooperative Action
BDEW	Bundesverband der Energie- und Wasserwirtschaft (*German Association of Energy and Water Industries*)
BDI	Bundesverband der deutschen Industrie (*Federation of German Industries*)
BEE	Bundesverband Erneuerbare Energie e.V. (*German Renewable Energy Federation*)
BINGO	Business NGO
BSR	Business for Social Responsibility
CAN	Climate Action Network
CCOO	Confederación Sindical de Comisiones Obreras (Trade Union Confederation of Workers' Commissions, Spain)
CDP	Carbon Disclosure Project
CDU	Christian Democratic Union (Germany)
CEP	Communications, Energy and Paperworkers Union of Canada
CEO	Chief Executive Officer
CFDT	Confédération Française Démocratique du Travail (French Democratic Federation of Labour)
CFMEU	Construction, Forestry, Mining and Energy Union (Australia)
CGT	Confédération Générale du Travail (General Confederation of Labour, France)

CIEL	Center for International Environmental Law
CJA	Climate Justice Alliance
CLC	Canadian Labour Congress
CO_2	Carbon Dioxide
COGTA	Department of Cooperative Government and Traditional Affairs (South Africa)
COP	Conference of the Parties
COSATU	Congress of South African Trade Unions
CPRS	Carbon Pollution Reduction Scheme (Australia)
CSC	Confédération des Syndicats Chrétiens (Confederation of Christian Trade Unions, Belgium
CSD	Commission on Sustainable Development
CSP	Concentrated Solar Plant
CSU	Christian Social Union (Germany)
DGB	Deutscher Gewerkschaftsbund (*German Trade Union Confederation*)
DM	Deutschmark (former standard monetary unit of Germany)
DOE	Department of Energy
EFO	Fernández Oro Station (Argentina)
ENGO	Environmental NGO
ETU	Electrical Trades Union (Australia)
ETUC	European Trade Union Confederation
EU	European Union
FDP	Freie Demokratische Partei (*Liberal Democratic Party, Germany*)
GCCA	Global Call for Climate Action
GDP	Gross Domestic Product
GDR	German Democratic Republic
GHG	Greenhouse Gas
GTLC	Gippsland Trades and Labour Council (Australia)
GW	Gigawatt
ICC	International Chamber of Commerce
ICEM	International Federation of Chemical, Energy, Mine and General Workers' Unions
ICFTU	International Confederation of Free Trade Unions
IG BCE	Industriegewerkschaft Bergbau, Chemie, Energie (*Mining, Chemical and Energy Industrial Union, Germany*)
ILO	International Labour Organization
IMATU	Independent Municipal and Allied Trade Union (South Africa)

INDC	Intended Nationally Determined Contribution
IOE	International Organisation of Employers
IPCC	Intergovernmental Panel on Climate Change
IPP	Independent Power Producers (South Africa)
IRP	Integrated Resource Plan (South Africa)
ISTAS	Instituto Sindical de Trabajo, Ambiente y Salud (*Trade Union Institute of Work, Environment and Health, Spain*)
ITUC	International Trade Union Confederation
IWGIA	International Working Group for Indigenous Affairs
JTA	Just Transition Alliance (United States)
JTC	Just Transition Centre
LMBV	Lausitzer und Mitteldeutsche Bergbau-Verwaltungsgesellschaft (*Lusatian and Central German Mining Management Company*)
LVA	Latrobe Valley Authority (Australia)
MCA	Minerals Council of Australia
MW	Megawatt
Mt	Megatonnes
MXGM	Malcolm X Grassroots Movement
NDC	Nationally Determined Contribution
NDP	New Democratic Party (Canada)
NERSA	National Energy Regulator of South Africa
NGO	Non-governmental Organisation
NUMSA	National Union of Metalworkers of South Africa
OCAW	Oil, Chemical and Atomic Workers' Union (United States)
OECD	Organisation for Economic Co-operation and Development
PACJA	Pan African Climate Justice Alliance
PAE(G)	Pan American Energy (Group)
PC	Progressive Conservative (Canada)
PCB	Polychlorinated Biphenyl
PPA	Power Purchase Agreement
PSI	Public Services International
R&D	Research and Development
REACH	Registration, Evaluation, Authorisation and Restriction of Chemicals – Regulation [EC] No 1907/2006
RED	Regional Electricity Distributor
REFIT scheme	Renewable Energy Through a Feed-in Tariff
REIPPPP	Renewable Energy Independent Power Producers Procurement Programme

RINGO	Research NGO
SAMWU	South African Municipal Workers Union
SEC	State Electricity of Victoria (Australia)
SPD	Sozialdemokratische Partei Deutschlands (*Social Democrat Party of Germany*)
SSEG	Small-scale Embedded Generation
TUAC	Trade Union Advisory Council (to the OECD)
TUC	Trades Union Congress (United Kingdom)
TUCA	Trade Union Confederation of the Americas
TUED	Trade Unions for Energy Democracy
TUNGO	Trade Union Non-governmental Organisation
UK	United Kingdom
UMW	United Mine Workers (United States)
UN	United Nations
UNEP	United Nations Environment Programme
UNFCCC	United Nations Framework Convention on Climate Change
UnTER	Unión de Trabajadoras y Trabajadores de la Educación de Río Negro (*Union of Education Workers of Rio Negro*)
US	United States
USW	United Steelworkers (United States)
VER.DI	Vereinte Dienstleistungsgewerkschaft (*United Services Trade Union, Germany*)
WBCSD	World Business Council for Sustainable Development
WCL	World Confederation of Labour
WEDO	Women's Environment and Development Organization
WTS	Worker Transfer Scheme (Australia)
WWF	World Wide Fund for Nature
YPF	Yacimientos Petrolíferos Fiscales (*Fiscal Oilfields – Argentine energy company*)

Acknowledgments

This book builds on many years of research on labour environmentalism and stakeholder dynamics in international climate and development debates. It also draws on more recent exchanges and debates within the Just Transition Research Collaborative (JTRC). Launched in early 2018, the JTRC brings together approximately 20 activist scholars from the Global North and South who have a shared interest in the just transition concept and narrative, and, more broadly, the social justice dimensions of the transition toward a low-carbon world. We wish to thank all those involved in the JTRC for their ideas and support. Special thanks go out to Diego Azzi, Jacklyn Cock, Romain Felli, Nicole Helmerich, Josua Mata, Hadrian Mertins-Kirkwood, Tadzio Müller, Sandra van Nickerk, Vivian Price, Rebecca Shelton and Damian White who actively took part in the Collaborative's activities over the past year and worked on the report 'Mapping Just Transition(s) to a Low Carbon World', which was launched at the Katowice Climate Conference in December 2018. We also wish to thank John Barry and Sinéad Mercier, Thomas Hirsch, Peter Newell and Anabella Rosemberg for their noteworthy contributions to the JTRC's online forum. Their and JTRC members' analyses and insights fuelled our own thinking on just transitions and greatly contributed to making this book a reality. We gratefully acknowledge funding for JTRC activities from the Rosa-Luxemburg-Stiftung (RLS) with support from the German Ministry for Economic Cooperation and Development, and thank Till Bender and Ethan Earle in particular for the support they extended to the JTRC.

We wish to thank all the contributors to this volume who provided excellent contributions and timely responses to our requests. Finally, we're grateful to the publishing team at Pluto Press – in particular, David Castle, Robert Webb and Melanie Patrick – and the manuscript's copy-editor, Jeanne Brady, for their impeccable work.

Edouard, Dunja and Dimitris

Introduction: The genealogy and contemporary politics of just transitions

Dimitris Stevis, Edouard Morena & Dunja Krause

LESSONS FROM KATOWICE

Organised in early December 2018 in the heart of Polish coal country, the Katowice Climate Conference (COP24) was billed the 'Just Transition COP' by participants and observers. As Kate Wheeling explains in the *Pacific Standard*, it 'was meant to be the one that prioritized the rights and needs of workers whose livelihoods are dependent on fossil fuels, so that they don't suffer as countries work to decarbonize their economies' (Wheeling, 2018). Given its symbolic location, COP24 was seen as an opportunity to focus the international community's attention on the justice and equity dimensions of climate mitigation and adaptation. It was a chance to counter a resurgent 'jobs vs environment' discourse and hopefully make progress in an international climate negotiation space that was struggling to deliver on the Paris Climate Agreement, especially following Donald Trump's decision in 2017 to exit the agreement. The hope for the United Nations Framework Convention on Climate Change (UNFCCC) – recently rebranded as UN Climate Change – and the host country was to use COP24 to generate 'momentum' and send strong 'signals' that the low-carbon transition was not only good for the climate and the economy, but good for workers and communities as well – and especially those whose livelihoods still depended on coal.

On the face of it, COP24 appears to have delivered on its promise. As part of the official conference, an 'Ambition and Just Transition Day' (10 December 2018) was organised and over 25 side events were devoted to the issue (Jenkins, 2019). These were opportunities for stakeholders to launch and showcase reports and initiatives, and share stories of just transitions in action. The Polish presidency of the COP, with support from around fifty governments, drafted and launched the 'Solidarity and Just Transition Silesia Declaration'. In the document, signatories 'stress that just transition of the workforce and the creation of decent work and quality jobs are crucial to ensure an effective and inclusive transition to low greenhouse gas emission and climate resilient development'.[1] A series of just transition-themed events were also organised on the margins of the official conference space. These included the Climate Hub, a civil society space hosted by Greenpeace, where a number of talks and presentations were organised on just transition. These and other just transition-related efforts within and around COP24 were not new but a product of the concept's mainstreaming in the UN space over the past decade, in particular following its inclusion in the preamble of the Paris Agreement in 2015. The Agreement refers to the need to '[take] into account the imperatives of a just transition of the workforce and the creation of decent work and quality jobs in accordance with nationally defined development priorities' (UNFCCC, 2015).

Yet, on closer scrutiny, the 'Just Transition COP', rather than providing a clear sense of how a just transition can be achieved, exposed the gap between climate policy makers' narrow understandings of just transition, and the complex and multifaceted reality of a 'living concept' whose origins and meanings lie deep in the everyday experiences of workers and frontline communities. It also exposed the gap between governmental endorsement of just transition and the reality on the ground; a reality in which the most vulnerable sections of society and least responsible for the climate crisis are either made to pay the price for low-carbon transition or used/manipulated to justify climate inaction or low ambition.

The fact, for instance, that the host country, Poland, presented itself as a champion of just transition while simultaneously backing the coal industry and calling for an 'evolutionary transformation of the power sector, instead of drastic restriction on the use of fossil fuels' (Darby, 2018), is revealing of this gap between discourse and reality. The conference organisation and venue also embodied this contradiction. The conference's main sponsors were the state-owned coking coal company, Jastrzebska Spólka Weglowa SA, the state-owned utility company Polska Grupa Energetyczna SA, which burns more coal than any other power company in Europe, and the Katowice-based Tauron Polska Energia SA, which owns coal mines. If this was not enough, the host city's official booth featured large cages full of coal and household products made of ash (Chemnick, 2018).

As the climate conference unfolded, two major events contributed to emphasise the importance of a just transition in addressing the justice and equity dimensions of climate change, while further highlighting the multiple and at times contradictory approaches to it. France, whose

---

1   https://cop24.gov.pl/fileadmin/user_upload/Solidarity_and_Just_Transition_Silesia_Declaration_2_.pdf

government had officially endorsed the Silesia Declaration[2] and whose President, Emmanuel Macron, was designated 'Champion of the Earth' by UN Environment in 2018, was in the midst of what would become the greatest social movement protest since May 1968. Just two days before the Katowice COP, 'Yellow Vest' protesters hit the global headlines during a day of violent clashes with the police in the streets of central Paris. Images of burning vehicles, barricades and a ransacked Arc de Triomphe were all over the news. The movement was triggered by a tax increase on fuel whose proceeds were intended to fund the low-carbon energy transition. The 'Yellow Vests' expressed, among other things, a growing sense of anger at the fact that the country's increasingly cash-strapped and precarious working and lower-middle class, was unjustly being made to pay for the energy transition. The movement highlighted the gap between political rhetoric and the reality of French climate policies. At a press conference on 4 December 2018 in Katowice and in response to the French government's decision to freeze the tax increase following the preceding day of protests, Pierre Cannet of WWF France summed up the situation in the following manner:

> There's no viable solution to reducing emissions on the scale needed in France without a price on carbon pollution as well as complementary policies, but a process that is not developed in an inclusive manner is destined to fail. Today's announcement that the French government is freezing carbon tax shows they put the cart before the horse by not addressing the social measures necessary for a just transition. Achieving decarbonization at the speed called for by science requires political will, and equity needs to remain at the core of the discussion. (WWF, 2018)

At around the same time and on the back of the 2018 mid-term elections, the equity and justice dimensions of the low-carbon transition also came to occupy the political debate in the United States. Just weeks before the COP, on 13 November, a group of activists from the Sunrise Movement, a youth-led political movement on the left of the Democratic Party, staged a sit-in in the office of Nancy Pelosi, the House Speaker, to get her to endorse a Green New Deal. This marked the beginning of a sustained campaign to persuade Congress to pass a ten-year plan to transition the United States towards a low-carbon economy. In February 2019, the newly elected Democratic Representative, Alexandria Ocasio-

---

2 https://cop24.gov.pl/fileadmin/user_upload/files/The_List_of_Leaders_and_
Parties_endorsing_the_Solidarity_and_Just_Transition_Silesia_Declaration.pdf

Cortez and Senator Ed Markey, presented a joint 'Green New Deal Resolution' (House Resolution 109, 2019) that explicitly referred to the need 'to achieve net-zero gas emissions through a fair and just transition for all communities and workers'. Growing calls for a Green New Deal have spurred a massive debate within the Democratic Party, as well as within and between labour and environmental justice groups – such as the Climate Justice Alliance (CJA) – that actively campaign for a just transition. In particular, it has led to interesting discussions on who should drive the just transition, and for whom.

ONE CONCEPT – DIFFERENT MEANINGS

What the Yellow Vest movement and Green New Deal debates highlight is that the Paris Agreement alone was not responsible for the popularisation of just transition. It is an idea that is currently being promoted by a range of actors. It reflects a growing awareness of and concern about deepening inequalities between the world's rich and poor, and how the climate and environmental crises, and efforts to address them, are accentuating them. The climate justice issue is increasingly being framed as one that cuts across national borders. The tension is more and more between a minority of super-rich individuals with high-carbon lifestyles, and a mass of poor people who are least responsible for the climate crisis but suffer the most from its effects *and* are disproportionately made to pay for climate mitigation and adaptation measures. As Oxfam showed in a report published just before COP21 in December 2015, the richest 10 per cent are responsible for almost 50 per cent of lifestyle consumption emissions, as opposed to the poorest 50 per cent, who are responsible for only about 10 per cent of lifestyle consumption emissions (Oxfam, 2015). In short, the notion of 'common but differentiated responsibilities' does not only apply between countries but within countries as well.

The current political climate marked by growing defiance towards political elites, and the ensuing resurgence of populist, xenophobic, nationalist, anti-climate, 'jobs vs environment' discourses – from Hungary to the United States, to Brazil and the Philippines – has also done a lot to raise just transition's profile. In particular, Donald Trump's victory in the 2016 US presidential election, following a campaign where he expressed his love affair with coal – 'Trump digs coal' – acted as a wake-up call for mainstream climate advocates. On the back of Trump's decision to exit the Paris Agreement, and in a bid to better integrate the social justice dimensions of the shift from a dirty to a green economy, various mainstream climate NGOs, think tanks, business interests and foundations – from Bloomberg Philanthropies to Greenpeace, to We

Mean Business and the World Resources Institute – appropriated the just transition concept.

The growing references to just transition undoubtedly signal a desire to further root social and equity concerns into the climate debate. While this is to be welcomed, it also complicates the task of identifying what just transition stands for, who is behind it, what are the underlying politics, and who it is for. Instead of leading to an alignment of views, the concept's growing popularity has actually turned it into a contested concept, like sustainable development (Connelly, 2007). It has created the conditions for struggles to impose a given understanding of what just transition should *actually* mean. What underlying theories of change and worldviews are associated with these various understandings of just transition? Are they mutually exclusive or compatible? Given the concept's growing popularity and centrality in the climate governance space – especially among corporate interests whose commitment to social justice is questionable to say the least – addressing these and other questions is essential for anyone who takes climate justice seriously. This appropriation of the concept is especially of concern to groups that were actively mobilising behind it long before it became fashionable in main-stream climate circles. As Jacqueline Patterson, director of the National Association for the Advancement of Colored People's Environmental and Climate Program explains, 'It's a concern when Big Greens and others are using the term and getting funded for using the term. It's become the term du jour for foundations, and those front-line communities become objectified' (McKibben, 2017).

The inclusion of just transition in the Paris Agreement, while raising awareness of the social dimensions of the low-carbon transition and contributing to the concept's popularisation, has also paradoxically con-tributed to de-historicise it and to – conveniently? – separate it from the frontline communities and labour unions that originally developed it and continue to mobilise around it in their day-to-day struggles. The risk is in losing sight of the fact that current references to just transi-tion in the Silesia Declaration, Paris Agreement and Green New Deal are the outcome of four decades of debates, campaigns and hard-won struggles by workers and frontline communities at the local, national and international levels. By paying lip service to the concept's history and embeddedness in the labour movement and frontline communities, policy makers and climate specialists 'uproot' the concept and empty it of its transformative, emancipatory and subversive potential that essen-tially comes from the fact that the just transition is both aspirational and grounded in people's everyday lives and struggles. As various contribu-tions to this volume highlight, just transition acts as a beacon to guide collective action and simultaneously gives rise to tangible alternatives on the ground.

This process of 'de-historicisation' and 'uprooting' of the concept through appropriation is not restricted to stakeholders in the climate debate but also characterises certain academic analyses. Indeed, as we will see in the following section, a significant part of the burgeoning academic literature tends to present just transition as a rigid, ahistorical concept for policy makers, and to downplay its essential function as a mobilising tool for the disenfranchised. Much of the research pays lip service to the decisive and historic role of labour and frontline groups in conceiving, nurturing and developing the concept over the past forty years, as well as their motivations for doing so.

JUST TRANSITION IN ACADEMIA

While academic research on just transition is fairly recent, there is a long history of applied and programmatic research on the topic, for the most part conducted by activists and organic intellectuals from within the ranks of the labour movement and associated groups and organisations. Union-friendly publications such as *New Solutions* in the United States – launched by union leader Tony Mazzocchi and his networks in 1990 (see below) – as well as the UK-based *Hazards* magazine, played an important role in hosting analyses and debates on the strategy of just transition (Slatin, 2002, 2009; Bennett, 1999). Rather than providing an analysis of this work here, we feel that it is more relevant to integrate it into the historical account that follows so as to better grasp its embeddedness and function for just transition advocates.

In the early 2000s, a handful of academic articles referred to the just transition concept. Noteworthy examples include an article on labour environmentalism in the United States (Gould et al., 2004), another on Australia with a focus on coal (Evans, 2007) as well as reflections by people directly involved (Bennett, 2007). Towards the beginning of the present decade, a larger body of academic research – in the field of labour environmentalism (Räthzel & Uzzell, 2013) – began to more systematically refer to and, on fewer occasions, focus on just transition (Räthzel et al., 2010; Snell & Fairbrother, 2011; Stevis, 2011, 2013). It is also worth highlighting the precursory role of a small group of Australian and South African academics (Cock, 2011, 2015; Snell & Fairbrother, 2011, 2013; Goods 2013).

While most early research was rooted in the experience and politics of the labour movement, a 2012 volume by Mark Swilling and Eve Annecke, entitled *Just Transitions: Explorations of Sustainability in an*

Unfair World presages a different approach that combines sociotechnical transitions with social justice and transitional justice approaches – in this case in South Africa (Swilling & Annecke, 2012). This has led Peter Newell and Dustin Mulvaney (2013) to argue in their much-cited article that '[in] academic circles the [just transition] term derives from a set of literatures on "socio-technical" transitions ... that are also increasingly being applied to questions of energy politics and policy' (Newell & Mulvaney, 2013:133). While some of the more recent research on just transition may in fact draw from the field of sociotechnical transitions, such an affirmation does not do justice to the fact that most just transition research was and has been grounded in labour environmentalism and, more problematically, contributes to obscure the concept's origins. Instead of sociotechnical transitions, a more appropriate connection would have been with sociotechnical systems, an approach developed after the Second World War to deal with the design of work in UK coal mines but which is not acknowledged by sociotechnical transition analysts (Cohen-Rosenthal, 1997; Cohen-Rosenthal et al., 1998). Additionally, just sustainability transitions have more affinity with just transition (Agyeman et al., 2016).

Swilling and Annecke's book fits into a first category of literature that focuses on broad topics or issues, such as sustainable development, justice, or energy transitions, but tends to leave out or downplay the role of organised labour or other specific actors. This body of work refers to just transition in a rather generic manner and with little to no references to the concept's history or to research – especially from labour environmentalism – that adopts a historical approach (Swilling & Annecke, 2012; Jasanoff, 2018; Heffron & McCauley, 2018). Within this category, some analyses, especially on energy transitions and environmental justice, do include limited references to the concept's origins and history (Farrell 2012; Newell & Mulvaney, 2013; Routledge et al., 2018). Such research can contribute to a broader understanding of just transition provided it does not strip it of historical agency – whether unions or other social forces – and does not downplay the importance of the research and analysis produced by unions and their allies over the years.

A second body of work recognises the importance of specific historical actors and relations, including labour. Within this category, a first subcategory specifically looks at the labour origins of just transition and explores union and union-ally efforts to implement just transition policies (Snell & Fairbrother, 2013; Goods, 2013; Felli & Stevis, 2014; Hampton, 2015; Snell, 2018; Morena, 2018). Drawing extensively on Felli's (2014) work on global union environmentalism, Dimitris Stevis and Romain Felli, for instance, explore the variability of global unions' approaches to just transition (Felli, 2014; Stevis & Felli, 2015).

A second subcategory recognises the origins of just transition and the centrality of unions but also situates just transition within broader political and theoretical debates (Cock, 2011; Stevis & Felli, 2016; Evans & Phelan, 2016; Healy & Barry, 2017; JTRC, 2018; White, 2019; Ciplet & Harrison, 2019). This engaged research combines historical depth and contemporary analysis of both the just transition concept, as well as its links to broader transformations on the left (Barca, 2015a). In a recent article, and drawing on the current Green New Deal debates in the United States, Damian White offers a fascinating analysis of the possible convergences between just transitions and 'design for transitions' currents. In particular, he looks at how they could 'facilitate modes of antiracist, feminist and ecosocialist design futuring that can get us to think beyond degrowth/Left ecomodern binaries and toward a design politics that can support a Green New Deal' (White, 2019:1).

The just transition concept's growing popularity has led some academics to categorise the uses and understandings of just transition (Cock, 2011, 2015; Felli & Stevis, 2014; Stevis & Felli, 2015; Hampton, 2015; JTRC, 2018; Barca, 2015b). In their recent article, David Ciplet and Jill Lindsey Harrison, for instance, focus on the different understandings of just transition and the concept's increasingly contested nature (see also Goddard & Farrelly, 2018; Barca, 2015a). Having retraced its evolution and appropriation by environmental justice groups, they explore 'existing and potential areas of conflict, tensions, and trade offs within just transition planning and activism' that derive from this (Ciplet & Harrison, 2019:1). Stefania Barca shows how just transition demands range 'from a simple claim for jobs creation in the green economy, to a radical critique of capitalism and refusal of market solutions' (Barca, 2015b: 392).

As our brief – and necessarily incomplete – overview of the literature indicates, many academic studies of just transition either leave out or only mention the concept's origins in US labour environmentalism and its globalisation through the agency of national and global labour unions and environmental justice groups. They tend, and this is justifiable given the urgency of the climate crisis, to focus on its contemporary uses, as well as its conceptual and theoretical implications and potential. Yet, by downplaying the importance of the concept's history and the centrality of agency, they run the risk of downgrading – and even omitting – grassroots and labour contributions to debates around low-carbon transitions and further reinforcing the misleading narrative that labour and nature do not mix (for a view of the breadth of labour environmentalism based on theory and cases, see Räthzel & Uzzell, 2013). In the following section, we seek to rectify this by providing a comprehensive history of just transition. This, we believe, is essential to fully grasp the similarities and contradictions between the different uses of the just transition concept in the different chapters of this book.

THE EMERGENCE OF JUST TRANSITION: 1980S–C.2001

Just transition was not the product of theoretical debates over environmental justice or sociotechnical transitions. It was developed during the 1970s and 1980s by workers in response to 'job blackmail' from capital and its allies under the increasingly hyperliberal capitalist turn unfolding in the United States.[3] Over time it expanded both geographically and to other constituencies through the efforts of national and global labour unions. Unlike various other concepts that have spread throughout the global environmental or developmental field (such as 'sustainable development' or 'green growth'), just transition's emergence was geographically and socially rooted.

The idea behind what was eventually called 'just transition' was born in the United States, in the 1970s. Most observers agree that it was the brainchild of Tony Mazzocchi – a trade unionist working on occupational safety and health at the Oil, Chemical and Atomic Workers' Union (OCAW). Just transition, while not initially referred to in those terms, was the product of his determined efforts to reconcile environmental and social concerns. As far back as the 1950s, Mazzocchi had been exposed to social environmentalist ideas. Unlike many fellow trade unionists, his priority was not to make all jobs safer. He acknowledged that certain jobs were too detrimental to workers, society and the environment, and should therefore be scrapped – in the case of nuclear weapons for instance – or replaced by automation (Leopold, 2007). In all cases, the priority should be to empower workers and communities, and enable them 'to know and act', especially in the face of job blackmail. This, he believed, could be achieved through the strengthening of labour environmentalism. As early as 1973, for example, Mazzocchi successfully enlisted support from environmentalists to help OCAW wage what he presented as 'the first environmental strike' over health and safety issues at Shell refineries across four US states.

---

3   This strategic connection should not obscure the fact that just transitions are not limited to the environment and that transitional strategies in response to various other transitions, such as offshoring or demobilisation, have been common if not satisfactory.

Mazzocchi was not operating in a vacuum. Other unions had also shown an interest in the intersections between jobs and the environment and had supported environmental legislation early on. In 1976, for instance, the United Automobile Workers organised a conference entitled 'Working for Economic and Environmental Justice and Jobs' (Bryant, 1997; Rector, 2014, 2018). The conference brought together people from various demographics as well as local and national unions and environmentalists, who shared a broad understanding of nature-society relations. A group called Environmentalists for Full Employment also sought for a number of years to bring together environmentalists and unions and published what can be considered as the first green industrial policy proposal that addressed work and workers (Grossman & Daneker, 1979).

The rise of hyperliberalism during the 1980s allowed corporations to engage in job blackmail, as well as to actually move highly toxic operations off shore (Leonard, 1988), leading to a precipitous drop in the OCAW's membership. For a while, the union merely reacted to this general trend because its resources were tied up in the five-and-a-half year strike (June 1984 to December 1989) against chemicals giant BASF (Minchin, 2002). International political shifts allowed for a resurgence of labour environmentalism within the union. In 1987, it received a major grant to develop a model health and training programme for workers. In 1989, it called for a single payer national health care programme for the United States and at its 1991 convention it 'passed a resolution calling for "A New Social, Political, and Economic Agenda" which set goals for the 1990s, including national health care, a Labor Party alternative, environmental protection, a Superfund for Workers, and international trade unionism'.[4]

The Superfund for Workers was the first name for just transition. It was developed by Mazzocchi (1993) and a network of activists – including Mike Merrill and Les Leopold (for a broader discussion, see Wykle et al., 1991), and was a direct response to the dominant 'jobs versus environment' discourse (OCAW, 1991), a discourse that was fuelled by the neoconservative right in response to environmental regulation (see, for example, Gollop & Roberts, 1983). In the same way that earlier evocations of a GI Bill for Workers sought to capture the imagination of the public, the Superfund for Workers evoked the Superfund programme that was put in place in 1980 at the federal level to fund the clean-up of thousands of contaminated industrial sites – manufacturing facilities,

---

4   Information on developments between 1987 and 1991 from www.usw-608. com/ocaw-history-page.html. See also Wykle et al., 1991.

processing plants, landfills and mining sites – where hazardous wastes had been dumped.

It was in 1995 that Les Leopold and Brian Kohler first publicly referred to the just transition concept during a presentation to the International Joint Commission on Great Lakes Water Quality (Hampton, 2015). In their words:

> We propose that a special fund be established; a just-transition fund which we've called in the past a superfund for workers. Essentially this fund will provide the following: full wages and benefits until the worker retires or until he or she finds a comparable job; two – up to four years of tuition stipends to attend vocational schools or colleges plus full income while in school; three – post-educational stipends or subsidies if no jobs at comparable wages are available after graduation; and four – relocation assistance. (Leopold, 1995:83)

The breadth of OCAW's vision was evident in its role in catalysing the creation of the Just Transition Alliance (JTA) in 1997 (Harvey, 2018; Labor Network for Sustainability & Strategic Practice: Grassroots Policy Project, 2016:7). The JTA brought together environmental and social justice organisations that represented the most vulnerable and marginalised populations in the United States. In addition to its participation in innovative training initiatives, the JTA was also involved in a number of specific local campaigns that sought to bring together workers and frontline communities to enhance unionisation and raise awareness about their exposure to anti-union and environmentally unjust practices. An important product of this effort was a training manual entitled 'A Just Transition for Jobs and the Environment' produced by the Public Health and Labor Institutes (2000) which were the vehicles through which the OCAW network sought to engage with environmentalists and communities.[5] The training tools were intimately connected to these local campaigns and drew upon a longer history of training workers and unions on occupational health and safety (Slatin, 2002, 2009).

The goal was to diffuse unionism within the environmental movement and environmentalism within the union movement. In short, the just transition initiatives of the 1990s did not only target public policy; they also sought to bridge the gaps between workers, environmentalists and communities, and to foster an alternative socioenvironmental politics that valued collective representation by workers and communities around principles of social and environmental justice. While largely

---

5   The network still exists within the United Steelworkers.

anthropocentric, it did not merely view nature as a means towards greater social equality. While motivated by the maldistribution of environmental harms, it did not seek to solve them by simply displacing the problem across space or time but sought to address the root causes of those harms, including by downsizing the production of toxics.

The OCAW had a direct influence on the Canadian labour movement as well as sections of the British labour movement working on occupational safety and health (see *Hazards* magazine). The Canadian Communications, Energy and Paperworkers Union of Canada (CEP) (part of which used to be the Canadian arm of OCAW) adopted a just transition resolution in 1996 and, in 1997, the OCAW followed suit. The Spanish Comisiones Obreras (CCOO) was also aware of these developments (Leopold, 1995). In 1999, the Canadian Labour Congress (CLC) adopted a just transition resolution and produced a fairly detailed programme of action in an ultimately unsuccessful effort to push Canada to adopt an ambitious climate policy (CLC, 2000; for more on the CLC's efforts, see Bennett, 2007).

The efforts of the OCAW and its allies were also reflected at the international level. In November 1999, for instance, the International Federation of Chemical, Energy, Mine and General Workers' Unions (ICEM), adopted a just transition resolution at its second world meeting. In the late 1990s and early 2000s, the International Confederation of Free Trade Unions (ICFTU) and the Trade Union Advisory Council (TUAC) to the Organisation for Economic Co-operation and Development (OECD) included just transition language in their occupational safety and health and environmental activities. The ICFTU's position at the climate conference (COP3) in Kyoto (1997), for example, included the declaration that 'workers will demand an equitable distribution of costs through "just transition" policies that include measures for equitable recovery of the economic and social costs of climate change programmes' (ICFTU, 1997:1). A 2001 report by Winston Gereluk and Lucien Royer, who coordinated labour at the global level, particularly the UN Commission for Sustainable Development (CSD), placed just transition at the heart of labour's sustainable development agenda (Gereluk & Royer, 2001).

In addition to the just transition efforts of the OCAW and the JTA there emerged a second thread of labour environmentalism during the late 1990s that also sought to include just transition. In 1997, despite the opposition of the United Mine Workers, the American Federation of Labor and Congress of Industrial Organizations (AFL-CIO) established a BlueGreen Working Group. After two years of deliberations, the group held a two-day meeting at AFL-CIO headquarters (14–15 April 1999) which was attended by the Congress of South African Trade Unions

(COSATU) as well as ICEM. While some more radical environmental organisations were involved that was not the case with environmental justice organisations. Moreover, some of the participants were more interested in preventing just transition policies, rather than advancing them. Bob Wages, the last president of the OCAW and by then vice president of the union into which the OCAW had merged, co-authored the follow-up memo that specified just transition as one of the working group's priorities.[6] The next major meeting of the group was scheduled for 11 September 2001 (9/11). The meeting never took place and the nationalist and conservative policies that followed complicated the dialogue between unions and environmentalists. Nonetheless, the report that emerged from these efforts, drafted by two technical advisers to the group (Barrett et al., 2002), did address just transition, particularly in light of the market mechanisms proposed to deal with climate change. This report can be considered as the first applied scholarship publication that sought to fuse a green industrial policy with just transition. Around that time, and due to strong opposition by the Mine Workers, just transition fell off the US union and environmental movements' agendas (for more on the period see Renner, 2000:59–71).

It is possible to distinguish three lines of green or environmental transitions by the end of this period. The first one was proposed by the OCAW and its allies in public health and environmental justice. While highlighting the detrimental impacts on workers, this group also focused on frontline communities. It was less about speaking in their name than about getting them on board and mobilising them. While it called upon companies to negotiate with unions and communities, it also focused on the role of the state in promoting a green industrial policy, and in which unions were expected to play a key role. Finally, and as the 1991 resolution indicates, the OCAW's just transition efforts were part of an ambitious set of policy proposals and efforts aimed at reorganising the US political economy. It is worth highlighting that the OCAW approach did not challenge growth per se; however, it did question certain trajectories of growth, even when this meant sacrificing jobs for the greater benefit of workers, society and the environment.

The OCAW approach contrasted with that of the BlueGreen Working Group which accepted the hegemony of market mechanisms and presented green economic growth as inherently positive for workers, especially given its job-creating potential (Barrett et al., 2002). Such an approach saw just transition as a corrective of the dislocations of green

---

6   Wages resigned from his new position in the spring of 2001 in reaction to the anti-environmental politics of the union into which the OCAW merged.

industrial policy. A third approach, emerging in the United States and elsewhere, left out just transition policies and placed all its hopes in green industrial policy (Apollo Alliance, 2004). This approach has historically dominated mainstream labour environmentalism in the United States but, as we will see shortly, it is not limited to the United States.

LABOUR AND THE GLOBALISATION OF JUST TRANSITION: 2001–2013

By 2001, the concept of just transition had gone into hibernation in the United States as well as globally, particularly due to opposition by the AFL-CIO (see Rosemberg, in this volume). By the end of the decade, however, just transition re-emerged nationally and internationally (Hampton, 2015). An important moment in this regard was the 2006 merger of the ICFTU and the World Confederation of Labour that gave birth to the International Trade Union Confederation (ITUC). From the outset, the ITUC placed environmental concerns at the heart of its agenda (Rosemberg, in this volume). Another noteworthy milestone includes the production of a policy brief in the lead-up to the 2009 Copenhagen Climate Conference (COP15) that explicitly linked the just transition concept to climate policy (ITUC, 2009a). Shortly after, an issue of the *International Journal of Labour Research* – the in-house journal of the International Labour Organization's Bureau for Workers Activities – was dedicated to just transition (Rosemberg, 2010a,b). Overall, the role of global union organisations, increasingly coordinated by Anabella Rosemberg of the ITUC, was paramount but behind them there was a network of national union pioneers.

The globalisation of just transition owes a great deal to the efforts of national unions. Noteworthy examples of unions that actively connected labour and environmental issues include the CCOO in Spain, the Trades Union Congress (TUC) in the United Kingdom, and the Australian Council of Trade Unions as well as the Australian Manufacturing Workers' Unions. While just transition was largely a strategy promoted by unions in the Global North, it did have some diffusion in the Global South. In South Africa, the National Union of Metalworkers of South Africa, for example, organised a training programme to spread environmentalism within its ranks and raise capacity to support a just transition (Cock, 2011).

Under the leadership of Joaquin Nieto and as far back as the late 1980s, the CCOO adopted a proactive approach to the environment and climate that was partly informed by efforts in the United States. In

1996, the CCOO established the Instituto Sindical de Trabajo, Ambiente y Salud (ISTAS),[7] which served as a research and education arm that would allow the union to integrate environmental and health and safety priorities into its practice, demands and collective agreements (Martin Murillo, 2013; Gil, 2013).

Another important actor in the diffusion of just transition was the British TUC. The TUC had focused on sustainable development and environmental issues since the late 1980s, and was actively involved in European and international climate debates. From 2005 to 2007, under the guidance of Philip Pearson, the TUC sponsored a research project on just transition conducted by the Working Lives Research Institute (Hampton, 2015), which resulted in a report on just transition (TUC, 2008). The TUC's participation in global climate negotiations also contributed to the diffusion of just transition with respect to climate policy.

Just transition was also being referred to by Australian unions who had a long history of labour environmentalism, going back to the green bans of the early 1970s (Burgmann & Burgmann, 1998). As an extractive economy, Australia faces significant challenges related to climate as well as pollution from lignite mining (Snell, in this volume; Goods, in this volume). Over the course of the 1990s, Australian unions explored the promise of green jobs (Crowley, 1999) and during the early years of the millennium Australian unions participated in and explored just transition solutions (AMWU, 2008; Evans, 2007; Snell & Fairbrother, 2011). By the end of this period South African unions were also engaged (e.g., COSATU 2011)

At the international level, a key promoter of just transition was the Sustainlabour Foundation, a green think tank closely linked to the research and educational arm of the CCOO and active at the international level (Martin Murillo, 2013; Hunter, 2010). Launched in 2004 and directed by Laura Martin Murillo, Sustainlabour – over its twelve years of existence – organised training sessions for union members, and published thematic reports, case studies, and policy recommendations, a number of which focused on developing contexts (see Rosemberg, in this volume; Martin Murillo, 2013). In close collaboration with the ITUC, it played a pivotal role in getting workers' voices heard in national and international policy spaces such as UN Environment (UNEP), the International Labour Organization (ILO), the UNFCCC and the CSD. Sustainlabour was also instrumental in organising the first Trade Union Assembly on Labour and the Environment in January 2006, which produced a landmark res-

---

7   At its height, ISTAS employed over a hundred people, including experts in occupational health and safety.

olution (UNEP, 2007, Annex 1). A second assembly was organised in 2012, in the lead-up to the Rio+20 Conference.

Drawing on the work of the 2006 Trade Union Assembly, the 2007 UNEP Labour and the Environment report explores the range of environmental transitions faced by labour and society at large, including climate change, toxics and health, production and consumption, and so on (UNEP, 2007). It provides an inclusive vision with respect to 'for whom' just transition should take place.

Given its growing importance, the UN climate process – through the UNFCCC – became a privileged venue for the ITUC, Sustainlabour and national unions to push a just transition agenda. As a result, just transition was increasingly framed and recognised as the trade union movement's contribution to the international climate debate. In a flyer produced in the lead-up to the 2009 Copenhagen Climate Conference, the ITUC presented just transition as 'a tool the trade union movement shares with the international community, aimed at smoothing the shift towards a more sustainable society and providing hope for the capacity of a "green economy" to sustain decent jobs and livelihoods for all' (ITUC, 2009b:1).

Trade unions that were engaged in the climate process pushed for the inclusion of just transition in successive UNFCCC decisions and agreements to highlight the benefits of decisive climate action for workers and their communities. Just transition also represented a way of mainstreaming environmental issues within the union movement and building bridges with other – especially environmentalist – actors engaged in the international climate debate (Morena, 2015; Rosemberg, in this volume). Building on the growing awareness and public concern for climate change and linking it up to the global economic crisis, the ITUC – as well as global union federations such as the International Transport Workers' Federation, Public Services International and IndustriALL – made a credible case for greater union engagement in the environmental field.

By shedding light on the social implications of climate change, just transition filled an important gap in the international climate debate. Until the early 2000s, equity and justice issues had been almost exclusively framed along a North-South axis. The priority for many climate justice activists involved in and around the UNFCCC had been to get developed countries to recognise their historical responsibilities for climate change and to act upon them – both through more ambitious national mitigation efforts and through higher levels of financial and technological assistance to developing countries (that are much more vulnerable to climate change). When climate justice groups referred to the uneven social impacts of climate change, they tended to focus on geographical differences (Fisher & Galli, 2015). Limited attention was paid to the differentiated social implications of both climate change and climate policies on the world of work in both the Global North and South (JTRC, 2018).[8]

For the ITUC, the aim of just transition was to 'strengthen the idea that environmental and social policies are not contradictory but, on the contrary, can reinforce each other' (Rosemberg, 2013:19). Through its efforts, especially in the lead-up to the Paris Climate Conference (COP21), the international trade union movement got certain UN agencies and programmes to adopt the just transition concept and language, contributing to its further diffusion within the international development and environmental community. Despite differences among them it is fair to say that global unions sought to integrate some kind of just transition within the political economy they were advancing (Felli, 2014; Stevis & Felli, 2015).

This is worth highlighting because the Great Recession accelerated the production of green growth and green capitalism proposals (Jacobs, 2012; Tienhaara, 2014), some including just transition as a crisis management tool and others not referring to it at all. The 2008 Green Jobs report published by UNEP, the ILO, IOE (International Organisation of Employers) and ITUC did include a chapter on just transition (Renner et al., 2008). However, the term is only mentioned once in the foreword and not at all in the sectoral chapters, leading one to think that it was added in response to political negotiations. Most importantly, the report associated just transition with decent employment and opportunities for employment, leaving out the policies necessary to transition specific workers and communities out of sunsetting industries, as envisioned in the original Superfund for Workers.

Other unions, particularly but not exclusively in the United States, were more attracted by proposals that avoided just transition altogether (Pollin et al., 2008, 2009). This was the case of the AFL-CIO, whose Energy Working Group avoided just transition terminology. This is not surprising since the president of the AFL-CIO at the time had previously been the president of the United Mineworkers during the 1990s.

It is worth noting that, unlike today, the various Green New Deal proposals at the time did not refer to the just transition concept, and in many cases, did not even focus on work and workers. While not referring

---

8   This and a few other passages in this introduction were adapted from the report *Mapping Just Transition(s) to a Low-Carbon World* published by the Just Transition Research Collaborative in November 2018.

to just transition, Green New Deal proposals by UNEP (Barbier, 2009; UNEP, 2009), the Green New Deal Group (2008), the German Greens (Bündnis 90/Die Grünen, 2012), or the Böll Foundation (Heinrich Böll Stiftung, 2009) did address ways in which to prepare workers for the green economy but without elaborating on questions of justice.

DIFFUSION BEYOND UNIONS: 2013 TO THE PRESENT

Just transition language has entered the mainstream of international, national and subnational climate-related debates. The concept is now used by UN and other intergovernmental organisations, governments, environmental and development NGOs, indigenous groups, feminist groups, businesses and philanthropists, to name a few. As has previously been highlighted, the term's growing popularity has led to a diversification of meanings associated with it. In the United States, for instance, there is a marked difference between national unions and grass-roots environmental and labour justice groups' approaches and understandings of just transition. Despite its growing popularity, we should not overstate the concept's significance. Many within the environmentalist and union spaces are still reluctant to employ it while there are cases of unions that were formerly leaders in the just transition debate that are now backtracking and abandoning it entirely. And even though just transition has gained traction in the international policy space and the Global North, it is still much less used or discussed in the Global South (see Cock, 2011, 2015; Kenfack, 2018; Satgar, 2018; Hirsch et al., 2017).

As was highlighted earlier, the previous decade was characterised by a growing, if tentative, institutionalisation of the just transition concept at the international level. Just transition, for instance, was explicitly referred to in the Green Jobs Initiative (2009–14), a joint initiative of UNEP, the ILO, the ITUC and the IOE. The initiative's goal was to encourage 'governments, employers and workers to collaborate on coherent policies and programmes to realise a sustainable and just transition with green jobs and decent work for all'.[9] Through this and similar UN-backed initiatives, just transition was increasingly placed on other global agendas and was also gradually associated with concepts such as 'green and decent jobs' (ILO) and 'green economy' (UNEP). The ILO's adoption of a just transition agenda in 2013 and just transition guidelines (ILO, 2015; International Labour Office, 2018) and, of course, the inclusion of just transition in the Paris Agreement (which was followed by the UNFCCC

---

9 www.ilo.org/beijing/what-we-do/projects/WCMS_182418/lang--en/index.htm

Secretariat commissioning a technical report on just transition from the ILO), have all contributed to popularise the concept in international and national policy circles. Other intergovernmental organisations are also taking an interest in the topic, but not always with a focus on workers and frontline communities, and are likely to use 'inclusive' rather than 'just' transition. In 2017, for instance, the OECD commissioned a report on just transition from the newly created Just Transition Centre (Smith, 2017), while the World Bank (2018) explored the equity challenges of a green transition in coal mining.

Its growing popularity among non-labour organisations notwith-standing, just transition remains firmly rooted in the union movement at the international level. The union movement's active presence in the international negotiation space, its sustained efforts to mainstream environmental and climate concerns within the union community, and its successful lobbying efforts to include just transition language in the 2015 Paris Agreement on climate change have also contributed to further anchor the concept within the union movement. At the international level, the ITUC and International Transport Workers' Federation have the most elaborate just transition policies (Stevis & Felli, 2015). Other global union organisations, such as IndustriALL (2019) and Public Services International, are also playing an increasingly active role in the just transition debate as has the Trade Union Network for Energy Democracy (Sweeney & Treat, 2018).

The 2016 launch of the Just Transition Centre (JTC) signals the beginning of a new phase in the ITUC's just transition efforts – one that reflects a renewed commitment to collaborative industrial relations. The centre's strategy centres around close collaboration with two global green business groupings – The B Team and We Mean Business – that are actively involved in the international climate arena. This collabora-tion between business interests and the JTC resulted in the publication of a *Just Transition: Business Guide* (JTC & The B Team, 2018). It is worth mentioning that the launch of the JTC coincided with the closure of Sustainlabour which, as we have shown, was a primary driver of labour environmentalism and just transition at the international level, moreover one with a strong Southern focus.

The global diffusion of just transition, particularly in the context of the international climate negotiations, has led to the adoption of just transition language by the most prominent environmental NGOs and networks. The reference to just transition in the preamble of the Paris Agreement further legitimised the concept and encouraged a wider range of stakeholders to use it. This was complemented by the concept's compatibility with the agreement's voluntary and bottom-up approach and the wider narrative on the combined economic, social and environ-mental benefits of climate action, especially in the energy field (Aykut et al., 2017; Morena, 2016). Many stakeholders now refer to just transition in their campaigns and publications (Nelson, 2018; The Lofoten Decla-ration, 2017). These include the Sierra Club in the United States, as well Greenpeace and Friends of the Earth International. The largest global network of NGOs working on climate change, Climate Action Network International, has also taken an interest in the concept and developed advocacy positions on the topic (CAN, 2018).

A number of regional, national and subnational governments – the European Union, Canada and Spain, Scotland, New York State, Colorado for example – have suggested or developed just transition task forces, commissions, funds, or policies. Most of these efforts aim at the tran-sition out of coal (see chapters by van Niekerk, Snell, Reitzenstein et al., and Mertins-Kirkwood & Hussey in this volume). This focus, while important, has contributed to narrowing down the scope of just transi-tion and to leaving out other sectors, as well as workers and communities that will be affected by climate change and efforts to address it (*Interna-tional Journal of Labour Research*, 2014).

In parallel, just transition has also made a noteworthy comeback in the United States. At the grass-roots level, community-based labour and environmental justice organisations and networks are actively cam-paigning for a just transition that is not restricted to labour issues or dirty energy, but also focuses on cultural, gender and racial injustices and is connected to a more general critique of extractive capitalism. Noteworthy examples include the Climate Justice Alliance and its 'Our Power: Communities for a Just Transition' campaign (Climate Justice Alliance, 2017a; Akuno, in this volume). As Ciplet and Harrison explain:

in practice, [environmental justice] and just transition organiza-tions, networks, and movements in the United States have extended the frame of a just transition broadly to issues facing communities including environmental racism, zero waste, energy democracy, mass incarceration and inequitable policing, gentrification, and Indigenous rights and sovereignty, among others. (Ciplet & Harrison, 2019:4)

The US resurgence is interesting, given national unions' continued reluctance to fully engage in the just transition debate, and this despite repeated efforts from networks such as the Labor Network for Sustain-ability and the Climate Justice Alliance (2017b) to place just transition at the heart of the national union agenda. Yet, and while unions are no longer central, one could argue that, through their focus on workers and frontline communities, environmental justice groups that mobilise around the just transition are in fact reconnecting with its initial meanings and uses, as laid out by Tony Mazzocchi, the OCAW and the JTA in the 1990s (described above). In that respect, it is worth following the efforts of both the Climate Justice Alliance, that does not directly include unions but prioritises the role of workers and frontline communities, and the People's Climate Movement, that does include some unions as well as a range of societal organisations.

Other noteworthy and innovative initiatives come from the philanthropic sector. These include the Just Transition Fund in the United States, launched in April 2015 with support from the Rockefeller Family Fund and Chorus Foundation, whose mission is to support Appalachian coal-dependent communities to transition to a strong, resilient and diversified economy.[10] German political foundations have also been active in the international just transition space. More mainstream climate funders such as Bloomberg Philanthropies and the European Climate Foundation have also incorporated just transition wording into their work – for example, the Beyond Coal campaigns in the United States and Europe – as have foundations involved in the recently established F20 Platform. At a recent G20 event in Argentina, the platform called upon political leaders to take action for a just transition.

We are therefore currently witnessing an accelerating diffusion of just transition across social forces and organisations. The variability of just transitions is impressive if we were to map them in terms of frequencies. However, it is still possible to identify those that place just transition within a broader and more cohesive political economy, those that see it as a corrective of the tensions of green transitions and, finally, those that put their hopes in the positive impacts of green growth. Just transition's globalisation and diffusion has not ended the tensions among these visions. Rather, it has made them more apparent and ubiquitous.

LEARNING FROM THE PAST TO FIGHT FOR THE FUTURE

Our overview of the origins and trajectory of just transition has allowed us to reposition the world of labour and frontline communities in their rightful place in relation to the environmental debate. It has more broadly involved reaffirming the view that meaningful social or environmental change cannot happen without the emancipation of workers. This does not mean essentialising workers by overstating their role, or placing them above other alienated and oppressed groups, but rather of

---

10   www.justtransitionfund.org/

highlighting that without profound changes in the organisation of and relations at work a just, inclusive and transformative socioenvironmental transition is impossible. For too long, and despite the efforts of certain environmental historians (Dewey, 1998; Montrie, 2008; Barca, 2016; Rector, 2014) and environmental labour studies researchers (Räthzel & Uzzell, 2013), the academic and policy debate has been dominated by the 'jobs vs environment' binary and the idea that labour and other 'old social movements' were indifferent or opposed to environmental concerns.

Historicising just transition is also important because it allows us to show that the concept was not always associated with climate or energy (and more narrowly, coal) policy – something that is widely assumed, especially following its inclusion in the preamble of the Paris Agreement. As we have shown, over the course of the 1980s and 1990s, just transition was mostly associated with toxics and their impacts on workers, and the local environment. In short, just transition is about more than responding to the energy crisis. It is also about addressing a broader range of issues, from the low-carbon transition to the uses and abuses of ecosystems, to our food systems (UNEP, 2007; *International Journal of Labour Research*, 2014).

Finally, historicising the just transition concept also enables us to highlight its multiple functions for workers and organised labour. Just transition is not only about creating a safety net for precarious workers in declining sectors. As various chapters in this book highlight, it is about getting workers and frontline communities to engage in the environmental debate, organising workers in ascending green sectors and getting those still dependent on polluting industries and sectors to accept and support a green transition (Álvarez Mullaly et al., Rosemberg, van Niekerk, Reitzenstein et al., all in this volume).

The most important historical lesson from the study of just transition is to avoid thinking of the last thirty years as an inexorable road from less to greater ambition. As we pointed out in the beginning, the proliferation of just transition references is evidence of the success of progressive labour. But it also carries with it the real potential of its appropriation by corporate interests for greenwashing purposes, and its manipulation by forces that are not interested in either green or just transitions (see the chapters of Moussu and Goods in this volume).

As Rector (2018) and Purdy (2018) suggest, the contemporary period lacks the strong democratic socialist vision that existed in the post-Second World War period and which informed the mobilisations of the 1960s and 1970s. The range of proposals and initiatives aiming at the empowerment of various constituencies is broader – as is the opposition

– but the overall political vision is less ambitious. While there have been spurts of hope in post-apartheid South Africa and post-dictatorship Brazil and, more recently in places such as Greece, Spain and the UK, the cause of a credible egalitarian and ecological left has been in decline. It is for this reason that the current traction of democratic socialism in parts of US society is so important, especially as such developments at the core of the world political economy have both domestic and global implications.

The current Green New Deal proposal supported by a number of US political leaders is an ambitious political proposal that does marry greening the economy with just transition. Its goal is to provide an agenda for the US left and it has increasingly been taken up as such in other parts of the world (Varoufakis & Adler, 2019; Alternative Information and Development Center, 2019). In a very real sense, it is both forward looking and at the same time reminiscent of Tony Mazzocchi's vision and the OCAW's just transition efforts, which in turn sought to revive the aspirations of the late 1960s and the 1970s before US hyperliberalism defeated the bastions of US democratic socialism. Historicising just transition therefore forces us to reflect on the broader historical trajectory in which it emerged and developed. When we do this, we realise that just transition involves more than simply helping workers to adapt to changes in their industries. It is about fulfilling a broader and deeper egalitarian and ecological vision (Álvarez Mullally et al., in this volume). The question becomes one of seeing whether contemporary just transition advocates will follow Mazzocchi's lead and foster a broad, counterhegemonic coalition of unions, environmentalists and justice advocates, or risk allowing this fascinating concept to be co-opted and stripped of its transformative potential.

REFERENCES

Agyeman, J, D Schlosberg, L Craven & C Matthews. 2016. 'Trends and directions in environmental justice: From inequity to everyday life, community and just sustainabilities', *Annual Review of Environment and Resources*, vol. 41, pp. 421–420.

Alternative Information and Development Center. 2019. *Understanding the Green New Deal.* http://aidc.org.za/understanding-green-new-deal/ (last accessed May 2019).

AMWU (Australian Manufacturing Workers' Union). 2008. *Making Our Future: Just Transitions for Climate Change Mitigation.* Granville, Australia: AMWU.

Apollo Alliance. 2004. *A Bold Program for Achieving America's Energy Independence.* Washington, DC.

Aykut, S, J Foyer & E Morena (eds). 2017. *Globalising the Climate: COP21 and the Climatisation of Global Debates*. London & New York: Routledge.

Barbier, E. 2009. *A Global Green New Deal. Executive Summary*. Report prepared for the Green Economy Initiative and the Division of Technology, Industry and Economics of the UN Environment Programme. www.edwardbbarbier. com/Publications/a_global_green_new_deal-executive_summary.pdf (last accessed April 2019).

Barca, S. 2016. 'Labor in the age of climate change: Any just transition to a green economy must take place on labor's terms – not capital's', *Jacobin*, 18 March. www.jacobinmag.com/2016/03/climate-labor-just-transition-green-jobs/ (last accessed April 2019).

——. 2015a. 'Labour and climate change: Towards an emancipatory ecological class consciousness', in L Temper & T Gilbertson (eds), *Refocusing Resistance to Climate Justice: COPing in, COPing out and Beyond Paris*, EJOLT Report 23, pp. 74–78. www.ejolt.org/2015/09/refocusing-resistance-climate-justice-coping-coping-beyond-paris/ (last accessed May 2019).

——. 2015b. 'Greening the job: Trade unions, climate change and the political ecology of labour', in RL Bryant (ed.), *International Handbook of Political Ecology*, pp. 387–400. London: Edward Elgar.

Barrett, JPJ, A Hoerner, S Bernow & B Dougherty. 2002. *Clean Energy and Jobs: A Comprehensive Approach to Climate Change and Energy Policy*. Washington, DC: Economic Policy Institute and the Center for a Sustainable Economy.

Bennett, D. 2007. 'Labour and the environment at the Canadian Labour Congress – The story of the convergence', *Just Labour: A Canadian Journal of Work and Society*, vol. 10, pp. 1–7.

——. 1999. 'Prevention and transition', *New Solutions: A Journal of Environmental and Occupational Health Policy*, vol. 9, no. 3, pp. 317–328.

Bryant, B. 1997. 'The role of the SNRE in the environmental justice movement', *University of Michigan School of Natural Resources and Environment*, 1 February. http://umich.edu/~snre492/history.html (last accessed April 2019).

Bündnis 90/Die Grünen. 2012. 'Green New Deal – The German perspective', *Green European Journal*, vol. 1, pp. 13–22.

Burgmann, M & V Burgmann. 1998. *Green Bans, Red Union: Environmental Activism and the New South Wales Builders Labourers' Federation*. Sydney: University of New South Wales Press.

CAN (Climate Action Network). 2018. *G20 Issue Brief: Just Transition*. http://climatenetwork.org/sites/default/files/can_g20_brief_2018_just_transition_1.pdf (last accessed May 2019).

CLC (Canadian Labour Congress). 2000. *Just Transition for Workers During Environmental Change*. Ottawa: Canadian Labour Congress. https://digital. library.yorku.ca/yul-1121737/just-transition-workers-during-environmental-change/datastream/OBJ/download (last accessed April 2019).

Chemnick, J. 2018. 'Poland celebrates coal as talks start in mining capital', *E&E News*, 6 December. www.eenews.net/stories/1060108909 (last accessed May 2019).

Ciplet, D & J Harrison. 2019. 'Transition tensions: Mapping conflicts in movements for a just and sustainable transition', *Environmental Politics*, published online, 31 March. https://doi.org/10.1080/09644016.2019.1595883.

Climate Justice Alliance. 2017a. *Our Power Communities: Just Transition Strategies in Place*. https://climatejusticealliance.org/workgroup/our-power/ (last accessed May 2019).

——. 2017b. *Just Transition: A Framework for Change*. https://climatejusticealliance. org/just-transition/ (last accessed May 2019).

Cock, J. 2015. *Alternative Conceptions of a 'Just Transition' from Fossil Fuel Capitalism*. Johannesburg: Rosa-Luxemburg-Stiftung. www.rosalux.co.za/ wp-content/uploads/2015/10/JCock-Futures-Commission-3_2015.pdf (last accessed January 2019).

——. 2011. 'Contesting a "just transition" to a low carbon economy', *Global Labour Column*, no. 76. University of Witwatersrand. http://column.global-labour-university.org/2011/01/contesting-just-transition-to-low.html (last accessed May 2019).

Cohen-Rosenthal, E. 1997. 'Sociotechnical systems and unions: Nicety or necessity', *Human Relations*, vol. 50, no. 5, pp. 585–604.

Cohen-Rosenthal, E, B Fabens & T McGalliard. 1998. 'Labor and climate change: Dilemmas and solution', *New Solutions: A Journal of Environmental and Occupational Health Policy*, vol. 8, no. 3, pp. 343–363.

COSATU (Congress of South African Trade Unions). 2011. 'A Just Transition to a Low-carbon and Climate Resilient Economy: COSATU Policy on Climate Change: A Call to Action'. COSATU.

Connelly, S. 2007. 'Mapping sustainable development as a contested concept', *Local Environment*, vol. 12, no. 3, pp. 259–278.

Crowley, K. 1999. 'Jobs and environment: The "double-dividend" of ecological modernisation?' *International Journal of Social Economics*, vol. 26, no. 7/8/9, pp. 1013–1027.

Darby, M. 2018. 'Polish government split over coal ahead of UN climate summit', *Climate Home News*, 21 November. www.climatechangenews.com/2018/11/21/ polish-government-split-coal-ahead-un-climate-summit/ (last accessed May 2019).

Dewey, S. 1998. 'Working for the environment: Organized labor and the origins of environmentalism in the United States, 1948–1970', *Environmental History*, vol. 3, no. 1, pp. 45–63.

Evans, G. 2007. 'A just transition from coal to renewable energy in the Hunter Valley of New South Wales, Australia', *International Journal of Environment, Workplace and Employment*, vol. 3, no. 3–4, pp. 175–194.

Evans, G & L Phelan. 2016. 'Transition to a post-carbon society: Linking environmental justice and just transition discourses', *Energy Policy*, vol. 99, pp. 329–339.

Farrell, C. 2012 'A just transition: Lessons learned from the environmental justice movement. *Duke Forum for Law and Social Change* vol.4, pp. 45–63.

Felli, R. 2014. 'An alternative socio-ecological strategy? International trade unions' engagement with climate change', *Review of International Political Economy*, vol. 21, no. 2, pp. 372–398.

Felli, R & D Stevis. 2014. 'La stratégie syndicale d'une "transition juste" vers une économie durable', *Mouvements*, vol. 80, no. 4, pp. 111–118.

Fisher, DR & AM Galli. 2015. 'Civil society', in K Bäckstrand & E Lövbrand (eds), *Research Handbook on Climate Governance*, pp. 297–308. Cheltenham: Edward Elgar.

Gereluk, W & L Royer. 2001. *Sustainable Development of the Global Economy: A Trade Union Perspective*. Geneva: International Labour Office.

Gil, BMT. 2013. 'Moving towards eco-unionism: Reflecting the Spanish experience', in N Räthzel & D Uzzell (eds), *Trade Unions in the Green Economy: Working for the Environment*, pp. 64–77. London: Routledge.

Goddard, G & M Farrelly. 2018. 'Just transition management: Balancing just outcomes with just processes in Australian renewable energy transitions' *Applied Energy*, vol. 225, pp. 110–123.

Gollop, FM & MJ Roberts. 1983. 'Environmental regulations and productivity growth: The case of fossil-fueled electric power generation', *Journal of Political Economy*, vol. 91, no. 4, pp. 654–674.

Goods, C. 2013. 'A just transition to a green economy: Evaluating the response of Australian unions.' *Australian Bulletin of Labour*, vol. 39, no. 2, pp. 13–33.

Gould, KA, TL Lewis & JT Roberts. 2004. 'Blue-green coalitions: Constraints and possibilities in the post 9-11 political environment', *Journal of World-Systems Research*, vol. 10, no. 1, pp. 91–116.

Green New Deal Group (on behalf of New Economy Foundation). 2008. *A Green New Deal: Joined-Up Policies to Solve the Triple Crunch of the Credit Crisis, Climate Change and High Oil Prices*. https://neweconomics.org/2008/07/green-new-deal (last accessed April 2019).

Grossman, R & G Daneker. 1979. *Energy, Jobs and the Economy*. Boston, MA: Alyson.

Hampton, P. 2015. *Workers and Trade Unions for Climate Solidarity: Tackling Climate Change in a Neoliberal World*. New York: Routledge.

Harvey, S. 2018. 'Leave no worker behind: Will the just transition movement survive mainstream adoption?' *Earth Island Journal*, Summer 2018. www.earthisland.org/journal/index.php/magazine/entry/leave_no_worker_behind/ (last accessed May 2019).

Healy, N & J Barry. 2017. 'Politicizing energy justice and energy system transitions: fossil fuel divestment and a "just transition"', *Energy Policy*, vol. 108, pp. 451–459.

Heffron, R & D McCauley. 2018. 'What is the "just transition"?' *Geoforum*, vol. 88, pp. 74–77.

Heinrich-Böll-Stiftung. 2009. *Toward a Transatlantic Green New Deal: Tackling the Climate and Economic Crises*. Publication Series on Ecology, vol. 3, prepared by the Worldwatch Institute. Brussels: Heinrich-Böll-Stiftung. www.boell.de/sites/default/files/green_new_deal.pdf (last accessed April 2019).

Hirsch, T, M Matthess & J Fünfgelt (eds). 2017. *Guiding Principles & Lessons Learnt for a Just Energy Transition in the Global South.* Friedrich-Ebert-Stiftung. https://library.fes.de/pdf-files/iez/13955.pdf (last accessed May 2019).

House Resolution 109. 2019. *Recognizing the Duty of the Federal Government to Create a Green New Deal.* 116th Congress. www.congress.gov/bill/116th-congress/house-resolution/109 (last accessed May 2019).

Hunter, C. 2010. 'Labor unions and sustainable development: Lessons from Sustainlabour', MA Professional Paper, Department of Political Science, Colorado State University.

ICFTU (International Confederation of Free Trade Unions). 1997. *Climate Change and Jobs: Towards a Strategy for Sustainable Development: Trade Union Statement to the Kyoto Conference (1–10 December 1997).* Brussels: ICFTU.

ILO (International Labour Organization). 2015. *Guidelines for a Just Transition Towards Environmentally Sustainable Economies and Societies for All.* Geneva: ILO.

IndustriALL. 2019. *A Just Transition for Workers: A Trade Union Guide.* http://www.industriall-union.org/sites/default/files/uploads/documents/Just_Transition/a_just_transition_-_english.pdf (last accessed June 2019).

International Labour Office. 2018. *Just Transition Towards Environmentally Sustainable Economies And Societies for All – ILO ACTRAV Policy Brief.* Written by Béla Galgóczi, Senior Researcher at the European Trade Union Institute (ETUI). At www.ilo.org/wcmsp5/groups/public/---ed_dialogue/---actrav/documents/publication/wcms_647648.pdf

International Journal of Labour Research. 2014. *A Just Transition for All: Can the Past Inform the Future?* www.ilo.org/wcmsp5/groups/public/---ed_dialogue/---actrav/documents/publication/wcms_375223.pdf (last accessed May 2019).

ITUC (International Trade Union Confederation). 2009a. *Trade Unions and Climate Change: Equity, Justice & Solidarity in the Fight Against Climate Change.* Brussels: ITUC. http://www.ituc-csi.org/IMG/pdf/climat_EN_Final.pdf (last accessed April 2019).

——. 2009b. *A Just Transition: A Fair Pathway to Protect the Climate.* Brussels: ITUC. www.ituc-csi.org/IMG/pdf/01-Depliant-Transition5.pdf (last accessed May 2019).

Jacobs, M. 2012. *Green Growth: Economic Theory and Political Discourse.* Center for Climate Change Economic and Policy Working Paper 108 and Grantham Research Institute on Climate Change and the Environment Working Paper 92. www.lse.ac.uk/GranthamInstitute/wp-content/uploads/2012/10/WP92-green-growth-economic-theory-political-discourse.pdf (last accessed May 2019).

Jasanoff, S. 2018. 'Just transitions: A humble approach to global energy futures', *Energy Research & Social Science*, vol. 35, pp. 11–14.

Jenkins, KEH. 2019. *Implementing Just Transition after COP24.* Climate Strategies. https://climatestrategies.org/wp-content/uploads/2019/01/Implementing-Just-Transition-after-COP24_FINAL.pdf (last accessed May 2019).

JTC (Just Transition Centre) & The B Team. 2018. *Just Transition: A Business Guide*. www.ituc-csi.org/IMG/pdf/just_transition_-_a_business_guide.pdf (last accessed April 2019).

JTRC (Just Transition Research Collaborative). 2018. *Mapping Just Transition(s) to a Low-Carbon World*. Geneva: Rosa-Luxemburg-Stiftung, University of London Institute in Paris & United Nations Research Institute for Social Development. http://www.unrisd.org/jtrc-report2018 (last accessed April 2019).

Kenfack, CE. 2018. *Changing Environment, Just Transition and Job Creation: Perspectives from the South*. Serie Estudios e Investigaciones. CODESRIA, IDEAs and CLACSO. http://biblioteca.clacso.edu.ar/clacso/sur-sur/201808 10041959/Changing_environment.pdf (last accessed June 2019).

Labor Network for Sustainability & Strategic Practice: Grassroots Policy Project. 2016. *'Just Transition' – Just What Is It? An Analysis of Language, Strategies, and Projects*. www.labor4sustainability.org/wp-content/uploads/2016/07/ JustTransitionReport-FINAL.pdf (last accessed May 2019).

Leonard, HJ. 1988. *Pollution and the Struggle for the World's Product*. Cambridge: Cambridge University Press.

Leopold, L. 2007. *The Man Who Hated Work but Loved Labor: The Life and Times of Tony Mazzocchi*. White River Junction, NH: Chelsea Green Publishing Company.

——. 1995. Statement at the International Joint Commission's 1995 Biennial Meeting on Great Lakes Water Quality 'Our Lakes, Our Health, Our Future'. 22–25 September 1995, Duluth, Minnesota, pp. 80–84. https://legacyfiles.ijc.org/publications/C46.pdf (last accessed April 2019).

Martin Murillo, L. 2013. 'From sustainable development to a green and fair economy: Making the environment a trade union issue', in N Räthzel & D Uzzell (eds), *Trade Unions in the Green Economy: Working for the Environment*, pp. 29–40. London: Routledge.

Mazzocchi, T. 1993. 'An answer to the work-environment conflict?' *Green Left Weekly*, no. 114, 8 September. www.greenleft.org.au/content/answer-jobs-environment-conflict (last accessed June 2019).

McKibben, B. 2017. 'Climate justice is racial justice is gender justice. What is a "just transition" anyway? Bill McKibben asks Jacqueline Patterson, the director of the NAACP Environmental and Climate Justice Program', *Nation of Change*, 21 August. www.nationofchange.org/2017/08/21/climate-justice-racial-justice-gender-justice/ (last accessed May 2019).

Minchin, T. 2002. *Forging a Common Bond: Labor and Environmental Activism During the BASF Lockout*. Gainesville, FL: University of California Press.

Montrie, C. 2008. *Making a Living: Work and Environment in the United States*. Raleigh, NC: University of North Carolina Press.

Morena, E. 2018. 'Securing workers' rights in the transition to a low-carbon world: The just transition concept and its evolution' in S Duyck, S Jodoin & A Johl (eds), *Routledge Handbook of Human Rights and Climate Governance*, pp. 292–298. New York: Routledge.

——. 2016. *The Price of Climate Action: Philanthropic Foundations in the International Climate Debate*, London: Palgrave.

——. 2015. 'Les reconfigurations environnementales du syndicalisme: Construction de positions et stratégies globales', in J Foyer (ed.), *Regards Croisés sur Rio+20: La Modernisation Écologique à l'Épreuve*, pp. 235–258. Paris: CNRS Editions.

Nelson, G. 2018. 'NGO brief: Just transition in focus', *IISD SDG Knowledge Hub*, 31 May. http://sdg.iisd.org/commentary/policy-briefs/ngo-brief-just-transition-in-focus/ (last accessed May 2019).

Newell, P & D Mulvaney. 2013. 'The political economy of the "just transition"', *The Geographical Journal*, vol. 179, no. 2, pp. 132–140.

OCAW (Oil, Chemical and Atomic Workers' Union). 1991. *Understanding the Conflict between Jobs and the Environment. A Preliminary Discussion of the Superfund for Workers Concept.* Denver, CO: OCAW.

Oxfam. 2015. 'Extreme carbon inequality. Why the Paris climate deal must put the poorest, lowest emitting and most vulnerable people first', *Oxfam Media Briefing*, 2 December. www-cdn.oxfam.org/s3fs-public/file_attachments/mb-extreme-carbon-inequality-021215-en.pdf (last accessed May 2019).

Pollin, R, H Garrett-Peltier, J Heintz, & H Scharber. 2008. *Green Recovery: A Program to Create Jobs and Start Building a Low-Carbon Economy.* Amherst, MA: University of Massachusetts-Amherst, Department of Economics and Political Economy Research Institute (PERI).

Pollin, R, J Heintz & H Garrett-Peltier. 2009. *The Economic Benefits of Investing in Clean Energy: How the Economic Stimulus Program and New Legislation Can Boost U.S. Economic Growth and Employment.* Washington, DC: The Center for American Progress.

Public Health and Labor Institutes. 2000. *A Just Transition for Jobs and the Environment. Training Manual.* New York: The Public Health and Labor Institutes.

Purdy, J. 2018. 'The long environmental justice movement', *Ecology Law Quarterly*, vol. 44, no. 4, pp. 809–864.

Räthzel, N & D Uzzell (eds). 2013. *Trade Unions in the Green Economy. Working for the Environment.* London: Routledge.

Räthzel, N, D Uzzell & D Elliott. 2010. 'The Lucas Aerospace experience: Can unions become environmental innovators?' *Soundings*, vol. 46, pp. 76–87.

Rector, J. 2018. 'The spirit of Black Lake: Full employment, civil rights, and the forgotten early history of environmental justice', *Modern American History*, vol. 1, no. 1, pp. 45–66.

——. 2014. 'Environmental justice at work: The UAW, the war on cancer, and the right to equal protection from toxic hazards in postwar America', *Journal of American History*, vol. 101, no. 2, pp. 480–502.

Renner, M. 2000. *Working for the Environment: A Growing Source of Jobs.* Washington, DC: Worldwatch Institute.

Renner, M, S Sweeney & J Kubit. 2008. *Green Jobs: Towards Decent Work in a Sustainable, Low-Carbon World.* Nairobi: UNEP/ILO/IOE/ITUC. www.ilo.

org/wcmsp5/groups/public/@dgreports/@dcomm/documents/publication/ wcms_098504.pdf (last accessed April 2019).

Rosemberg, A (ed.). 2010a. 'Climate change and labour: The need for a "just transition"', *International Journal of Labour Research*, vol. 2, no. 2.

——. 2010b. 'Building a just transition: The linkages between climate change and employment', *International Journal of Labour Research*, vol. 2, no. 2, pp. 125–161.

——. 2013. 'Developing global environmental union policies through the ITUC', in N Räthzel & D Uzzell (eds), *Trade Unions in the Green Economy: Working for the Environment*, pp. 15–28. London: Routledge.

Routledge, P, A Cumbers & KD Derickson. 2018. 'States of just transition: Realising climate justice through and against the state', *Geoforum*, vol. 88, pp. 78–86.

Satgar, V. (ed). 2018. 'Trade union responses to climate change and just transition', *South African Labour Bulletin*, vol. 42, no. 3.

Slatin, C. 2009. *Environmental Unions: Labor and the Superfund*. Amityville, NY: Baywood Publishing Company.

——. 2002. 'Health and safety organizing: OCAW's Worker-to-Worker Health and Safety Training Program', *New Solutions: A Journal of Environmental and Occupational Health Policy*, vol. 11, no. 4, pp. 349–374.

Smith, S. 2017. *Just Transition. A Report for the OECD*. Brussels: Just Transition Centre. www.oecd.org/environment/cc/g20-climate/collapsecontents/Just-Transition-Centre-report-just-transition.pdf (last accessed May 2019).

Snell, D. 2018. '"Just transition"? Conceptual challenges meet stark reality in a "transitioning" coal region in Australia', *Globalizations*, vol. 15, no. 4, pp. 550–564.

Snell, D & P Fairbrother. 2013. 'Just transition and labour environmentalism in Australia', in N Räthzel & D Uzzell (eds), *Trade Unions in the Green Economy: Working for the Environment*, pp. 146–161. London: Routledge.

——. 2011. 'Towards a theory of union environmental politics: Unions and climate action in Australia', *Labor Studies Journal*, vol. 36, no. 1, pp. 83–103.

Stevis, D. 2013. 'Green jobs. Good jobs? Just jobs? US labour unions confront climate change', in N Räthzel & D Uzzell (eds), *Trade Unions in the Green Economy: Working for the Environment*, pp. 179–195. London: Routledge.

——. 2011. 'Unions and the environment: Pathways to global labor environmentalism.' *WorkingUSA*, vol. 14, no. 2, pp. 145–159.

Stevis, D & R Felli. 2016. 'Green transitions, just transitions? Broadening and deepening justice', *Kurswechsel*, vol. 3/2016, pp. 35–45.

——. 2015. 'Global labour unions and just transition to a green economy', *International Environmental Agreements: Politics, Law and Economics*, vol. 15, no. 1, pp. 29–43.

Sweeney, S & J Treat. 2018. *Trade Unions and Just Transition. The Search for a Transformative Politics*. TUED Working Paper No. 11. New York: Trade Unions for Energy Democracy, Rosa-Luxemburg-Stiftung – New York Office and Murphy Institute. www.rosalux-nyc.org/ wp-content/files_mf/ tuedworkingpaper11_web.pdf (last accessed 31 October 2018).

Swilling, M & E Annecke. 2012. *Just Transitions: Explorations of Sustainability in an Unfair World*. Claremont, CA: UCT Press.

The Lofoten Declaration. 2017. www.lofotendeclaration.org (last accessed May 2019).

Tienhaara, K. 2014. 'Varieties of green capitalism: Economy and environment in the wake of the global financial crisis', *Environmental Politics*, vol. 23, no. 2, pp. 187–204.

TUC (Trades Union Congress). 2008. *A Green and Fair Future: For a Just Transition to a Low Carbon Economy, Touchstone Pamphlet No. 3*. London: TUC. www.tuc.org.uk/sites/default/files/documents/greenfuture.pdf (last accessed May 2019).

UNEP (United Nations Environment Programme). 2009. *Global Green New Deal. Policy Brief.* Geneva: UNEP. https://wedocs.unep.org/bitstream/handle/20.500.11822/7903/A_Global_Green_New_Deal_Policy_Brief.pdf?sequence=3&amp%3BisAllowed= (last accessed April 2019).

——. 2007. *Labour and the Environment: A Natural Synergy.* https://wedocs.unep.org/bitstream/handle/20.500.11822/7448/-Labour%20and%20the%20Environment_%20A%20Natural%20Synergy-2007739.pdf?sequence=3&isAllowed=y (last accessed May 2019).

UNFCCC (United Nations Framework Convention on Climate Change). 2015. Adoption of the Paris Agreement. Twenty-first session of the Conference of the Parties. UN Doc. No. FCCC/CP/2015/L.9/Rev.1. 12 December.

Varoufakis, Y & D Adler. 2019. 'It's time for nations to unite around an international green new deal', *The Guardian*, 23 April. www.theguardian.com/commentisfree/2019/apr/23/international-green-new-deal-climate-change-global-response (last accessed May 2019).

Wheeling, K. 2018. 'At COP24, will climate action include basic protections for human rights?' *Pacific Standard*, 14 December. https://psmag.com/environment/cop24-will-climate-action-include-basic-protections-for-human-rights (last accessed April 2019).

White, DF. 2019. 'Just transitions/design for transitions: Preliminary notes on a design politics for a Green New Deal', *Capitalism Nature Socialism*, published online. https://doi.org/10.1080/10455752.2019.1583762 (last accessed June 2019).

World Bank. 2018. *Managing Coal Mine Closure: Achieving a Just Transition for All.* www.worldbank.org/en/topic/extractiveindustries/publication/managing-coal-mine-closure (last accessed April 2019).

WWF (World Wide Fund for Nature). 2018. 'WWF Response to tax freeze following fuel protest', *WWF*, 4 December. wwf.panda.org/?339670/WWF-Response-to-France-Fuel-Tax-Freeze (last accessed May 2019).

Wykle, L, W Morehouse & D Dembo. 1991. *Worker Empowerment in a Changing Economy: Jobs, Military Production and the Environment.* New York: The Apex Press.

1

'No jobs on a dead planet': The international trade union movement and just transition

Anabella Rosemberg

INTRODUCTION

Over the past 15 years, the international labour movement has succeeded in finding its place and imposing its demands in the international climate conversation. Tremendous progress has been made since the late 1990s, a period characterised by deep suspicion between environmentalists and trade unionists, as well as a general lack of understanding on the labour side of how the environmental and climate crises affected the very possibility of protecting workers' rights and securing new ones. The trade union movement's growing presence and involvement in the international climate space is correlated with the emergence and development of a concept: just transition. Particularly since the 2009 United Nations Climate Conference in Copenhagen (COP15), the concept has come to represent the trade union movement's contribution to the climate debate. A combination of factors relating to the political and economic context and internal trade union dynamics allowed a very diverse trade union constituency to embrace the concept, and with that, opened the way for trade unions to be perceived as a constructive, forward-looking force in the climate debate.

In this chapter, I will look at how the development of the just transition concept acted as the main conduit through which the international trade union movement went from a defensive, 'jobs vs environment' approach to climate action to a far more forward-looking and ambitious one. I discuss the different stages that led the international trade union movement to develop and adopt an ambitious position on climate change – a position that was not 'labour-centric' but integrated a broad climate justice approach that could appeal to a wider audience, while reassuring the trade union movement on its role in defending workers' interests. I will pay particular attention to the role of just transition in that evolution as well as the changing dynamics within the international trade union movement that enabled that positive shift to happen. I will also discuss the efforts that were deployed by trade unions – both internally and externally – to broaden the scope of support for just transition in the run-up to the Paris Climate Conference (COP21, 2015), as well as efforts to better align trade union positions with those of the international climate movement.

As a disclaimer, this chapter will not expand on efforts within the International Trade Union Confederation (ITUC) to develop the specific *content* of just transition policies and strategies. A number of other articles and reports address this issue (Rosemberg, 2017). This chapter will rather look at the rationale for using the concept and how the concept, in less than two decades, has come to represent, at least in the international climate space, the idea that social justice and equity must be at the heart of the low-carbon transition.

BUILDING THE FOUNDATIONS OF INTERNATIONAL UNION WORK ON THE ENVIRONMENT

In the early 2000s, the dominant narrative on the link between jobs and environment was a negative one: 'It's either Kyoto or jobs.' Sustainable development was the *terme du jour* (NBC News, 2005). When trade unions did get involved, as was the case at the 1992 Rio Conference on Sustainable Development, their efforts were essentially focused on strengthening the so-called 'social dimension' of sustainability. This defensive, 'and social too' approach logically limited the role and place of unions in global environmental and sustainability debates.

It was not until the early 2000s that the first attempts were made to articulate policy demands following the adoption of the Johannesburg Plan of Implementation – the outcome of the 2002 Earth Summit. If not marked by significant progress on the policy front, the Earth Summit forced the International Confederation of Free Trade Unions (ICFTU, the predecessor of the ITUC) to articulate positions on specific sustainability issues. There again, however, union contributions on issues like water or food continued to focus on their social aspects instead of the articulation between the social and the environmental. Nonetheless, from 2002 onwards, two concepts began to be increasingly circulated and referred to in international union discussions and policy documents; concepts that signalled a growing desire to bridge the environmental and social divide: green jobs and just transition.

The relatively low visibility of unions and union proposals in international processes reflected the limited interest and involvement of national trade union centres and union federations in international affairs and processes. There was an overall lack of understanding on how distant and abstract global issues and international processes – like sustainable development or climate change – were relevant to their day-to-day struggles in the workplace or national contexts. There were no real efforts to inform, train, or organise union representatives on sustainability and environmental issues. When unionists did take an interest in international processes, few of them had formal mandates from their unions to engage in advocacy or campaigning. At the international level, the fact that trade unions were rarely recognised as a distinct constituency in environment-related discussions – apart from the UN Commission on Sustainable Development (CSD) – further distanced national union centres from the international space.

Perhaps most importantly, trade unions at the national level did not view the environment or climate change as issues deserving union attention at the international level – nor, in many cases, at the national level for that matter. The ICFTU's largest affiliate, for instance, the American Federation of Labor and Congress of Industrial Organizations (AFL-CIO), blocked any effort to positively advocate on environment at the international level. This was the case at the ICFTU's Miyazaki Congress (2004). Due to AFL-CIO pressures, the ICFTU in its Congress resolution, while officially condemning the Bush Administration's rejection of the Kyoto Protocol, abstained from calling on the United States to re-enter it (ICFTU, 2004). The ICFTU was therefore unable to publicly endorse the only multilateral agreement – the Kyoto Protocol – committing governments to concrete emission reduction targets.

It is worth mentioning that a handful of unions *were* in favour of ambitious climate action. A noteworthy example was the Spanish CCOO (Confederación Sindical de Comisiones Obreras), and more particularly its confederal secretary, Joaquin Nieto. Since the 1980s, Nieto had gone to great lengths to include environmental prerogatives into his union's work, and to develop a loose international network of like-minded 'union environmentalists' to coordinate actions and weigh in on European and international debates. This core group of committed individuals would later play a central role in developing and coordinating international union efforts on climate change and the environment.

From 2004 onwards, a series of developments profoundly transformed the international trade union movement's outlook and strategy in relation to environmental issues. From a 'lowest common denominator approach', the movement began advocating both stronger climate ambition and a just transition for those who would be directly affected by the low-carbon transition. Most importantly, and reconnecting with the fundamentals of labour internationalism, the movement now increasingly prioritised global action and justice over national interests as expressed by governments and powerful national economic players. In other words, the union movement found its rightful place at the heart of the global struggle for climate justice.

CONNECTING UNIONS TO THE ENVIRONMENT: FINDING OUR 'DEMAND', OUR 'SPACE', AND GETTING UNIONS TO BELIEVE IN THEM

In 2006, there was a change in the architecture of the international trade union movement. The two largest international trade union bodies – the ICFTU and the World Confederation of Labour (WCL) – merged to create the ITUC. The new organisation brought in new affiliates, including the formerly communist-linked Confédération Générale du Travail (CGT) in France, GEFONT Nepal and CGT Argentina, which had previously been barred from joining the IFCTU and WCL. This transition period provided unions with an opportunity to reflect on the past and discuss the future role of international trade unionism in the face of new global challenges, including the environment and climate change. In other words, conditions were met to discuss the environmental crisis and its inclusion within the ITUC's mandate.

It is in this context that, in 2006, the ICFTU and WCL, in partnership with the recently created Sustainlabour foundation (2004, see below), co-hosted the first Trade Union Assembly on Labour and Environment in the UN Environment Programme's offices in Nairobi. The Assembly was attended by over two hundred unionists from across the globe who shared a common commitment to environmental and climate action, and, in most cases, worked on topics that were directly and indirectly related to the environment such as toxics, water, energy, rights at the workplace and corporate social responsibility. Attended by both the ICFTU and WCL General Secretaries, the Assembly signalled a historic evolution in the international union movement's approach to the environment – especially when compared to the Miyazaki Congress two years earlier (ICFTU, 2004). In its final resolution, the Assembly agreed on a series of objectives including the need:

> To take urgent action on climate change in support of the United Nations Framework Convention on Climate Change and its Kyoto Protocol; to develop new and additional agreements for both developed and developing countries, taking account of common but differentiated responsibilities; to anticipate and minimize the negative

effects and maximize the positive effects on employment of mitigation; and to ensure the participation of trade unions in decision-making on climate change strategies. (UNEP, 2007:118)

The 'union environmentalists' who attended the Assembly left it with a stronger political mandate, and a greater sense of recognition and confidence in their ability to build a collective international union voice on environmental issues.

Building on the work of the Assembly, environmental protection was included into the ITUC's constitution at its founding Congress in Vienna (ITUC, 2006). In preparation for the 2009 Copenhagen Climate Conference, Guy Ryder, the newly elected General Secretary (and current Director General of the International Labour Organization (ILO)), instructed staff to draft a policy platform centred on green jobs and just transition, which was presented as the flipside of climate ambition[1] (Martin Murillo, 2013). Henceforth, international trade union efforts on the environment would no longer be the remit of a small and marginalised group of individuals but formed part of the international union movement's core strategy. Nor, for that matter, would they be constrained by one union's veto power. From a 'lowest common denominator' approach, the international trade union movement moved towards a more ambitious and affirmative stance on environmental issues.

Building a common narrative: From green jobs to just transition

With the onset of the 2008 financial crisis, the ITUC was in a strong position to articulate a collective response that combined social and environmental concerns. By then, the organisation had already done a significant amount of work on green jobs and their potential in a wide range of sectors – most notably in energy and transport. The combined efforts of the ITUC, and a group of individuals in the ILO – who saw an opportunity for the oldest UN organisation to become relevant in the environmental discussion – resulted in the first Green Jobs Report (Renner et al., 2008). UNEP and the International Organisation of Employers (the ITUC's business counterpart in the ILO) officially endorsed the report.

---

1 This mandate led to the first quite elaborate trade union statement in the lead-up to COP13: www.ituc-csi.org/IMG/pdf/COP13_Statement.pdf (last accessed June 2019). Contrary to previous statements, this one was validated by the ITUC General Council, and therefore had the status of agreed trade union policy in December 2007.

Regrettably, the report did not receive the expected or deserved attention. This was partly due to the fact that, at the time, most of the global climate community's attention was fixated on McKinsey's marginal cost curves, which illustrated how cheap climate action was when compared to the quite onerous – at least in the short term – investment plan suggested by the union movement (Enkvist et al., 2007). It was also related to the fact that during the financial crisis, the austerity narrative won out. The priority in 2008 was to cut back on public spending and reduce consumer demand to absorb the financial costs of the bank bailouts. In most countries, the result was austerity, as well as stagnating and even shrinking wages. While the green jobs narrative was one of hope and a first positive connection between labour and the environment, the unfavourable economic context meant that it did not yield the expected policy outcomes.

It is also worth noting that the green jobs narrative – and the green growth and green economy narratives more generally – received a mixed reception from trade unions. Three main critiques surfaced during the 2008–12 period. They were especially salient in the context of the Rio+20 conference in 2012, where the green economy and green growth narratives held centre stage. The first critique came from more radical union activists who argued that the green jobs narrative upheld and even reinforced existing global power dynamics, especially between the Global North and South. In particular, they argued that only rich economies in the Global North would attract green investments.

A second group of unions was worried that green jobs policies would negatively impact workers in polluting sectors. Instead of 'green jobs', they preferred the term 'greener jobs', thereby shifting the focus to the greening of polluting industries, rather than their full-scale removal. This second set of critiques revealed the wider union movement's inability – at the time – to address the challenges faced by sectors negatively impacted by climate policies. The green jobs agenda, while critical for building a narrative of change and progress, insufficiently addressed these challenges.

The third critique came from unions who were worried that new green jobs would not lead to improved working conditions, including gender relations at work. The importance of aligning green jobs with the decent work agenda – i.e., 'jobs that are productive, provide adequate incomes and social protection, respect the rights of workers and give workers a say in decisions which will affect their lives' (ILO, 2012:6) – was critical. For unions, this meant reaching out and organising workers in the new green industries.

These criticisms aside, the international trade union movement's focus on green jobs did nevertheless create useful and fruitful connections between jobs and the environment – connections that would feed into the just transition work. Instead of being on the defensive, unions were beginning to imagine the kinds of jobs they wanted for the future – and in the process visualising what would and would not form part of that future. This evolution in the movement's overall outlook was especially palpable during the Second Trade Union Assembly on Labour and Environment in Rio de Janeiro in 2012 (in the context of the Rio+20 Conference). The Assembly's final resolution was a clear endorsement of the overall strategy to date, albeit more nuanced when it came to the way forward (ITUC et al., 2012). In the resolution, the Assembly called on the ITUC to better address concerns about the social impacts of environmental policies, and to ensure that its advocacy efforts balanced climate ambition with the interests of those whose jobs were most at risk. The time had come for expanding, strengthening and becoming more vocal on just transition.

Finding a place when you are not necessarily welcome: Engaging with the UNFCCC

In 2006, on the back of the first Trade Union Assembly, the newly created ITUC sent a fairly large delegation of unionists – coordinated by Lucien Royer, who was responsible for occupational health issues and informally in charge of sustainability – to the Nairobi Climate Conference (COP12). Despite organising a side event and having fruitful coordination meetings, the impact of the trade union delegation was close to zero, and this was not due to a lack of effort and motivation. The UNFCCC (United Nations Framework Convention on Climate Change) process and UN climate conferences are unique. They are very different, for instance, from sessions of the CSD, where unions were also quite active. In the CSD, there are no draft texts or intermediate meetings, and unions are officially invited to take part in discussions. Furthermore, the turnout at CSD meetings is usually quite low. They usually attract limited media attention and have a comparatively limited influence on national policy debates.

The UNFCCC was an altogether different story. Climate COPs attract thousands of delegates. While they have a limited access to the real negotiations, non-state actors – or constituencies – are authorised to make short statements at the end of plenaries, and draft negotiation texts are not discussed and finalised at the COP but prepared months in advance in the so-called 'subsidiary bodies'. The diplomats, UN bureaucrats, scientists, experts and environmentalists who attend climate COPs – and in some cases since the first COP in Berlin in 1995 – form a close-knit community of individuals who are traditionally wary of newcomers, especially when these newcomers carry new ideas that risk challenging the more established science-centric and 'apolitical' approach. The challenge for unions – and possibly the greatest challenge in relation to the UNFCCC – was therefore to 'find one's place' in a complex and rather unwelcoming environment.

These obstacles notwithstanding, the ITUC began exploring ways of influencing the negotiations. At the time, most Parties to the negotiation regarded union demands as too vague or disconnected from the existing negotiation agenda. And when relevant, they were seen as adding an unnecessary burden to an already complex and challenging process. In addition to this, the Climate Action Network (CAN), which was the only climate-focused NGO network in the UNFCCC at the time, had made it abundantly clear that it did not see unions as forming part of the climate movement. The other two officially recognised constituencies – businesses and researchers – were equally unwilling to accommodate union views. As a result, trade unions, while present at the COP, could not take the floor, submit positions to the Secretariat, or be 'seen' as a stakeholder in their own right.

Building on Agenda 21, which is supposed to guide stakeholder voices in UN environmental conventions, a procedural battle was launched by unions to get them recognised as an official UNFCCC 'constituency', just like the three existing 'official' constituencies: environmental NGOs (ENGOs), business NGOs (BINGOs), and research NGOs (RINGOs). After several attempts, the UNFCCC Secretariat finally consented to grant unions constituency status on condition that they 'prove' their commitment to climate action through the submission of reports over a three-year period (something that the business constituency was not asked to do).[2] It is worth highlighting that initial Party and constituency reactions were a combination of disbelief ('in my country, unions block action') and apprehension ('they will add more demands to an already crowded climate negotiations space').

Obtaining official constituency status was fundamental for ensuring that the voice and demands of unions could be heard in the formal nego-

---

2 Reports, presented over three years, needed to be structured in terms of the global scope of the representation, its periodicity (in particular in intermediary meetings such as the subsidiary bodies) and the content of the policies submitted, which had to be 'supportive' of climate ambition – with no further definition of what that meant. Those reports, and their structure was never an issue brought to the Parties, as the acceptance of constituencies is a Secretariat prerogative.

tiating space. Unions could now take the floor on every relevant agenda item at each COP and subsidiary body. They could formally voice their ideas and proposals, and in particular those relating to just transition, and more generally, their support for ambitious climate action.

For unions, the UNFCCC became a key battleground in which to highlight their commitment to climate action and push their demands. It soon became clear, however, that in order to be successful, the union movement had to produce a narrative capable of attracting widespread support – one that merged labour-focused demands with broader calls for ambitious climate action. It also meant identifying the key players in the negotiation, developing a well-planned advocacy strategy, and building alliances with other non-state actors. This was a major challenge given the union movement's limited human and financial resources. The ITUC climate team was composed of one person, myself (following Lucien Royer's retirement in 2008), and two Sustainlabour staff members.

Having a union-owned climate demand – just transition – and a space – the UNFCCC – in which to advocate it was important but not suffi-cient. The ITUC also had to find a way of getting its affiliates to adhere to an ambitious union position on climate change at both the national and international levels.

Finding a way to maintain ambition in a consensus-based interna-tional organisation

Given its consensus-based approach to decision making, the interna-tional trade union movement's positions traditionally tend to reflect the lowest common denominator. Had it stuck to this approach in the lead-up to Copenhagen, the ITUC would have probably struggled to officially recognise climate change as a problem (let alone a big problem). For those of us in the movement working on climate change, it was clear that, at the very minimum, our positions would have to be in line with the climate science. In the lead-up to Copenhagen, this meant – like most other non-state actors in the climate space – acknowledging and following the IPCC's recommendations in its Fourth Assessment Report that developed countries should reduce their emissions by 25–40 per cent by 2020 (when compared to 1990 levels) (IPCC, 2007). From a policy perspective, this represented a major evolution for the union movement. It also produced a number of reactions, especially from affiliates who wondered why unions should position themselves on issues that were already being addressed by others, and, in their view, not directly related to workers.

For the ITUC, it was essential to combine science-based climate ambition and social justice into our climate negotiations strategy in order for unions not to appear as defensive and self-interested stakeholders. Getting unions to accept this took time and caused some internal tensions. It marked a clear departure from how many affiliates habitually viewed the ITUC: as an international union convener rather than an actor in international affairs. The plan was therefore to reach out to affiliates and get them to adopt a fairly ambitious and 'solid' position at the international level, with the expectation that this would then feed into their positions at the national level as well, and dissuade them from simply aligning their position with those of their national governments or employers. For this to happen, the *spirit* of the conversation had to simultaneously stay in line with the climate science and UNFCCC principles, including common but differentiated responsibilities, and at the same time respond to more traditional union concerns.

The three following examples provide a good illustration of the multiple and complex dynamics at play within the labour movement around climate change and in the lead-up to the Copenhagen conference. The first relates to the international union movement's efforts to build an autonomous union voice in developing countries. The second example relates to the ITUC's engagement with its biggest affiliate, the AFL-CIO. And the third relates to efforts deployed by the European trade unions and the European Trade Union Confederation (ETUC), and how these fed into the international union strategy.[3] The challenge throughout was to strike the right balance between the protection of workers and their communities, and the overall need to drastically reduce GHG emissions.

Supporting Southern voices – And that means unleashing them, not controlling them

In 2004, the creation of Sustainlabour by Joaquin Nieto (of the CCOO) and Laura Martin Murillo marks a first key moment in the short history of international trade union involvement in the climate and environmental debate. By 2016, the year of its closure, the union-affiliated organisation had trained thousands of union leaders in over three hundred national union centres and federations on the connections between sustainability, climate and social justice. Sustainlabour's efforts, especially towards unions in the Global South, played an important part

---

3   These three cases do not do justice to other dynamics of great interest for the author, such as the conversations with Japanese trade union RENGO, discussions with the non-ITUC affiliate ACFTU (China) and the positive dynamics with Australian and Canadian trade unions, just to mention a few.

in raising the overall level of union ambition at the international climate policy level. Before 2006, Southern unions had practically no voice in the international environmental and climate debate. And this, even as many workers in the Global South were already having to deal with the impacts of environmental degradation and climate-related impacts, further aggravated North/South and domestic inequalities.

Through the efforts of its director, Laura Martin Murillo, and her team, Sustainlabour got Southern unions to take up the just transition concept, and this, even when environmental justice continued to be framed by many Southern governments as a secondary issue behind economic growth and development. Its efforts involved getting Southern unions to question their countries' carbon-intensive development models, while simultaneously getting Northern unions to recognise their own countries' historic responsibilities when it comes to climate change. By getting more Southern unions involved in the international climate space, Sustainlabour helped to develop a more balanced position in the climate negotiations – one that clearly recognises that the poorest, a majority of whom live in the Global South, are the primary victims of the current environmentally destructive and unjust economic system.

Broadly speaking, two groups of people led the discussions on the environment and just transition in Southern unions. The first group was composed of individuals who worked on international affairs in their unions. They were well versed in the procedures and dynamics built around UN negotiations. For the most part, they also analysed and engaged in the international space through a 'North vs South' lens. For them, to question emerging economies' commodity-focused and carbon-intensive development models was yet another expression of the Global North's imperialism and protectionism. Having said this, this group of individuals was instrumental in denouncing and challenging certain Northern union attempts to downplay industrialised countries' historical responsibilities. Holding developed countries to account was critical for ensuring a balanced union approach in the international climate arena, where otherwise, only Northern workers' interests would be defended.

The other group brought together committed trade unionists who, for a combination of personal and political reasons, were convinced of the need to involve Southern union voices into the international climate debate. Through their courage and imagination to envision a different development itinerary for their unions and countries, Joaquin Turco (Argentina), Massiel Figuereo (Dominican Republic), Josua Mata (The Philippines), Angelique Kipulu (Chad), and Kingsley Ofei Nkansah

(Ghana), to name a few, made sure that the ITUC's just transition efforts were relevant to the Global South.

Working with the AFL-CIO

The lead-up to Copenhagen was marked by a significant evolution in the official US trade union stance towards international climate change negotiations. From a lowest common denominator approach that reflected the views of a handful of vocal sectoral unions – notably the United Mine Workers (UMW) – the AFL-CIO's position evolved to a less obstructionist and even constructive one. A handful of committed AFL-CIO officials – including Joe Uelhein, Bob Baugh and later on Brad Markell – was behind this evolution. They could count on the support of a small but vocal group of unions. These included the United Steelworkers (USW), who were involved in the BlueGreen Alliance with 'Big Greens' such as the Sierra Club (Foster, 2010). The fact, however, that the national AFL-CIO leadership was still reluctant to engage in domestic climate policy efforts, produced a gap between what the AFL-CIO could support internationally and what it advocated at the national level. Arguing that it risked dividing the organisation, the AFL-CIO president, who was formerly the president of the coal miners' union, blocked attempts to organise an internal debate on climate change.

Regardless, the afore-mentioned union officials continued to actively engage in the UNFCCC process and contribute to the ITUC's climate efforts. Their hope was that by familiarising US unions with the international climate policy space, and getting them to regularly engage with other unions, the AFL-CIO leadership would ultimately change its position and play a more proactive role in the domestic climate debate. Just before Copenhagen, their unyielding efforts started to pay off. The AFL-CIO's leadership officially acknowledged that the ITUC's position in the climate talks – which was the de facto position of the union movement – should no longer be a 'lowest common denominator' position, but a middle-ground position between what unions *should* support in order to be in line with the climate science, and what they could *actually* support at the national level. Their efforts also paid off at the domestic level. The AFL-CIO's involvement in the co-drafting of entire sections on just transition as part of the Waxman-Markey Bill – or American Clean Energy and Security Act – for example, was heavily influenced by its exposure to the international union movement's climate efforts (American Clean Energy and Security Act, 2009).

The AFL-CIO's greater interest and involvement in the domestic and international climate debate did not signify that all of its affiliates were convinced of the need to push for ambitious climate action. The ultimate failure to adopt the American Clean Energy and Security Act exposed continued divisions within US labour. The absence of federal regulation and support for workers across sectors opened the door to plant-by-plant measures that would involve significant job losses without appropriate compensation measures, investments, or just transition plans. Equally, without a clear federal plan, companies could freely engage in environmentally destructive projects and, as in the case of the Keystone XL oil pipeline, 'buy' the short-term support of workers and unions through promises of employment and union-friendly contracts. This would become a source of important tensions within the US union movement between those who were convinced of the need for ambitious union action on climate, and those who continued to prioritise short-term work-related issues.

That being said, the increasing salience of the climate issue within the US progressive camp, the recognition of workers and just transition in the Paris Agreement, and the publication of the Just Transition Guidelines by the ILO (2015) did ultimately lead to the adoption of the first climate change resolution at the 2017 AFL-CIO convention. The resolution can be seen as a confirmation at the domestic level of principles that the AFL-CIO had been supporting at the international level for just under a decade (AFL-CIO, 2017).

How European trade unions lifted public interest as a driving principle for our climate work

In 2006, three years before Copenhagen, there was a major dispute between the European Trade Union Confederation (ETUC) and some of its affiliates over the adoption of the REACH regulation on chemical substances at the EU level (Registration, Evaluation, Authorisation and Restriction of Chemicals – Regulation [EC] No 1907/2006). The dispute ultimately resulted in the ETUC's adoption of an ambitious position that placed the general interest before the particular and strictly economic interests of the chemicals industry.

Throughout the REACH discussions, the ETUC faced strong opposition from the German Mining, Chemical and Energy Industrial Union (IG BCE), which acted as the de facto spokesperson for the powerful German chemical industry. Through the determined efforts of a handful of ETUC officials – Joel Decaillon, Sophie Dupressoir, Toni Musu, among others – and key confederations, the European trade union movement managed to put the public interest and environmental/health protection before the expected job losses within the chemicals sector (which were largely overestimated). This internal 'victory' over REACH generated impetus for European unions to push for more ambitious emissions targets, even if this meant going against some powerful sectoral unions and employer groups.

From a very early stage, the ETUC could count on the support of a group of European unions, including Comisiones Obreras (CCOO) in Spain, the Fédération Générale du Travail de Belgique, the Confédération des Syndicats Chrétiens (CSC) in Belgium, and the ETUC Secretariat, which backed the IPCC-inspired 25–40 per cent emission reduction target for the developed world and the idea that industrialised countries should take the lead in terms of emission reductions. As with chemicals, however, the ETUC's efforts on climate change were also met with opposition from certain national union confederations and sectoral federations which called on it to abandon its appeals for ambitious climate policies and to only focus on the just transition. In the post-Copenhagen years, the ETUC's efforts to get European trade unions to position themselves on EU climate-related directives were systematically met by European industrial trade union federation responses stating the contrary. Breaking a tacit ETUC prerogative along the way, energy or industrial unions publicly communicated their disapproval of ETUC positions to businesses in their sector and to EU institutions. Rather than weakening the ETUC's position, this actually contributed to highlight, especially in the eyes of non-union stakeholders, the ETUC's commitment to ambitious targets and willingness to play a proactive role in the European climate conversation.

While changes in the ETUC Executive Committee and an anti-climate drive from the Polish confederation Solidarność weakened its position between Copenhagen and Paris, had the ETUC not made these early efforts, it is unlikely that the ITUC would have been in a position to build a critical mass of trade unions to move the internal conversation forward – especially given US unions' reluctance to act (see above). The ETUC's efforts, as well as those of a dedicated group of national union centres, are all the more praiseworthy given the EU Commission's profoundly anti-social and neoliberal orientation.

In brief, out of this period of intense internal debates and power dynamics at different levels of the trade union movement emerged a comprehensive ITUC climate strategy and position – a position that recognised the need for ambitious and scientifically grounded climate action, and simultaneously highlighted the importance of securing a just transition for workers. When the 400-person-strong trade union delegation arrived in Copenhagen in December 2009, the ITUC had a clear mandate and was in a strong position to weigh in on the negotiations. In the midst of all the confusion and extreme tension that surrounded COP15 – especially following the circulation of a non-official draft agreement – trade unions succeeded in persuading negotiators to include just transition language in the draft text ('ensuring a just transition of the workforce that creates decent work and quality jobs').[4] While the leaked document and ensuing diplomatic crisis ultimately resulted in the collapse of the negotiations, the reference to the just transition in the official draft text served as a validation for the advocacy strategy taken by the ITUC on climate change.

CONNECTING OUR DEMANDS TO THOSE OF THE CLIMATE MOVEMENT

In the previous section, we looked at some of the internal processes and dynamics that enabled the international union movement to develop, agree to and deploy a credible position on climate change. Despite some 'wins', such as the reference to just transition in the 'official' (although ultimately ignored) negotiation text, the dominant narrative in Copenhagen was centred on the idea that 'technical fixes' and 'low-hanging fruit' were the key to solving the climate crisis. This dominant approach to climate action sharply contrasts with the underlying logic behind just transition, which involves a high level of planning and multi-stakeholder dialogue and engagement. In other words, although the negotiating process was not far from including a reference to just transition, the dominant approach to climate action that came out of Copenhagen did not provide any assurances that the low-carbon transition would be a socially just one for workers and communities – especially in the Global South.

Now that the international union movement had a shared approach, the priority became that of encouraging other stakeholders to focus on the social dimensions of climate action, and to hopefully shift the broader climate narrative in the lead-up to Paris. The Rio+20 Summit was an important milestone in this respect (see above). For the ITUC, it was an opportunity to connect its climate-related demands to a range of social justice concerns, such as the need for social protection, tax justice, or decent work. The three years separating Rio+20 (2012) and the Paris Climate Conference (2015) were crucial not only in ensuring inclusion of the ITUC's demands in the final climate Agreement but also in (1) getting other stakeholders to better integrate the social justice dimen-

---

4   The text discussed, part of the so-called Ad Hoc Working Group on Long-term Cooperative Action (AWG-LCA), would not be fully agreed in Copenhagen and was passed into the following COP, in Cancun (COP16), where the Just Transition language was indeed confirmed as part of decision 1.CP/16 – https://unfccc.int/resource/docs/2010/cop16/eng/07a01.pdf#page=2 (last accessed June 2019).

sions of climate action, and (2) better aligning national union positions to the international union message.

Broadening the spectrum of support for just transition

A few months after the Copenhagen collapse, the ITUC was invited to a gathering of influential NGOs in London (Vidal, 2009). The meeting was organised by the Global Call for Climate Action (GCCA), a broad umbrella coalition that unions had joined in the run-up to Copenhagen.[5] The sombre mood of the meeting contrasted with that of the ITUC for whom, despite the negotiation debacle, Copenhagen was a major step in its climate mobilisation effort. While unions and the rest of the climate movement were in diametrically different emotional spaces, the London meeting also pointed to the growing atmosphere of trust between them and receptiveness to each other's ideas. It is in this context that just transition became not just a means of raising union awareness and ambition on climate change, but also the basis for conversations with non-union actors on climate justice. Through these conversations, just transition went from being a union demand to one that continued to be associated with unions but was henceforth also supported and adopted by a majority of non-state actors in the UN climate space. This contributed to increasing the legitimacy and visibility of the demand and with this, the legitimacy and visibility of unions.

Three parallel developments help to illustrate the growing convergence and trust between the international union movement and other NGOs in the climate space. The first relates to the growing interactions at the national level. While this chapter focuses on the international level – rather than country-specific union efforts – it is still worth noting that in the lead-up to the Copenhagen conference, only a handful of national union centres were involved in climate coalitions – most notably Spain's Coalicion Clima, Belgium's Coalition Climat, Canada's Climate Action Network (CAN). After COP15, however, a significant number of unions formally joined their country's national or regional CAN chapter, and in certain cases, more politically engaged coalitions such as Climate Justice Now – CUT Brazil, Trade Union Confederation of the Americas (TUCA), SENTRO Philippines. For unions, joining a climate coalition was a practical way of positioning oneself on issues that they were com-

---

5   It is important to remember that since the creation of the trade union NGO (TUNGO) status in the UNFCCC and the policy set up for unions in the process, the collaboration with NGOs in this space remained limited and many still saw unions as an adversary rather than a potential ally.

fortable in supporting but were unable to mobilize around – such as adequate climate finance mechanisms to assist developing countries in their emission reductions efforts and to cover their adaptation needs, for example.[6] Union involvement also led a number of national and regional coalitions to adopt just transition as part of their demands in the lead-up to Paris, further contributing to popularising the concept.

As I show in the following section, at the international climate policy level, two other complementary developments – the build-up of a coalition around human rights in the UNFCCC, and active behind-the-scenes cooperation between the ITUC and a small group of influential NGOs – had a major impact on unions' status and influence in the international climate space pre-Paris.

Connecting just transition to other human rights

The union movement's efforts to obtain observer status in the UNFCCC inspired other Agenda 21 groups – women, indigenous peoples, youth – to do the same. These new 'major groups' whose demands were frequently regarded as too narrow and group specific – not only by governments, but by a number of more established climate NGOs – began to individually develop elements of what would later be labelled a 'rights-based approach' to climate policy. Throughout the negotiation process, many of their demands were used by Parties to undermine and delegitimise others. Unions, for example, were often told that if references were made to just transition and/or decent work, then human rights demands would 'bubble up' and 'clog' the entire negotiation process (Duyck et al., 2018).

In an effort to break these negative dynamics, unions joined and co-led a group of organisations – led by Sébastien Duyck – in an effort to consolidate demands from all those groups. From just transition to intergenerational justice, to human rights to gender equality, to indigenous peoples' rights to food sovereignty, the idea was no longer to oppose rights-based demands but to group them together and present them as a unified and coherent package. These coordinated efforts led a number of organisations[7] to draft and circulate a paragraph proposal to be included in the Paris Agreement:

---

6 Climate finance was one of three core union demands in the period from Copenhagen to Paris, at the same level as emission reductions and just transition.
7 Key organisations included CIEL (Center for International Environmental Law), WEDO (Women's Environment and Development Organization), the ITUC, CARE, AIPP (Asia Indigenous Peoples Pact), the IWGIA (International Working Group for Indigenous Affairs), among others. https://seors.unfccc.int/

This Agreement shall be implemented … ensuring the integrity and resilience of natural ecosystems and ensuring the respect, protection, promotion and fulfilment of human rights, including the rights of indigenous peoples; gender equality and the full and equal participation of women; intergenerational equity; a just transition of the workforce that creates decent work and quality jobs; and food security. (WEDO, 2015)

Getting Parties to adopt the paragraph was never going to be easy, especially as different countries supported different rights-based approaches. Switzerland, for instance, championed gender rights but opposed just transition. The United States, in contrast, championed just transition but had strong reservations about including human rights language into the Agreement. The coalition held out, and while it ultimately did not get the exact wording into the Agreement or expected placement – in the Preamble rather than the body of the text – it was nevertheless successful in getting just transition and key rights-related language into the final Agreement. Beyond the Agreement, the fact that all members adopted just transition and included it into their advocacy efforts, contributed to further legitimise the concept and the need for social justice in the international climate space. Just transition was no longer the demand of a single group – trade unions – but of all non-state actors. In turn, this effort also led to a strong mobilisation of unions on other partners' priorities.

Unusual partners

A third and less well-known development was instrumental in ensuring that just transition was taken up by most non-state actors in the negotiations. It was also crucial in getting unions to better understand and navigate the intricate, complex and chaotic world of climate negotiations. It was at the 2012 Climate Conference (COP18) in Doha, as the negotiations were wavering, that the ITUC representative (myself), and other negotiation heads from several prominent environmental and development NGOs – WWF, Oxfam, ActionAid, Greenpeace and Friends of the Earth – released a joint press statement outlining a litmus test for climate ambition and raising the alarm about the current state of negotiations (Cassegard & Thorn, 2017:39; WWF, 2012). The statement marked the start of a close and fruitful working relationship. It was, for its partic-

applications/seors/attachments/get_attachment?code=MB655WJoNWAMX-I36KR2DGFBFI6KCB4A1 (last accessed June 2019).

ipants and given the diversity of organisations represented, a valuable space for intelligence sharing and internal and external strategising. At the following COP in Warsaw (2013), for instance, and in response to the slow progress in the negotiations and to calls from Southern groups – Asia Pacific Forum on Women, Law and Development (APWLD) and Pan African Climate Justice Alliance (PACJA) – we successfully got our own organisations and most other non-state actors to join a symbolic 'walk out' from the Conference under the slogan 'polluters talk, we walk'.[8]

This small space of trust ensured unions were at the heart of the civil society conversation and acted as a valuable space in which to build better mutual understanding, discuss sensitive topics, and defuse potential conflicts. It enabled the ITUC to effectively counter anti-just transition narratives by mobilising support from very powerful allies, and pushed ITUC policies to be more ambitiously aligned with civil society demands. This also sent a powerful signal to the union movement: unions were now accepted and treated on an equal footing. They were now at the heart of the climate space. Following the Warsaw COP, unions would systematically be consulted and integrated into civil society both within and at the margins of the official conference space.

When the once minority position on climate became mainstream in the union movement

At the ITUC leadership level as well, efforts were made to convince other stakeholders of the need for a just transition and inclusion of social justice and equity issues in the Paris negotiations. Since being elected General Secretary of the ITUC in 2010, Sharan Burrow, in her regular high-level meetings with heads of state and global climate leaders, deployed every effort to both highlight trade unions' commitment to ambitious climate action and convince other stakeholders of the need to address the social dimensions of the low-carbon transition.

Internally, she also made efforts to consolidate the progress that had already been achieved in the climate domain. During her mandate, climate change became one of three priority issue areas for the ITUC – along with ending slavery and securing a universal social protection floor. A first internal milestone was reached in 2010, at the ITUC's 2nd World Congress in Vancouver, where unions not only discussed their role in the climate debate, and the means to ensure a just transition towards

8   www.theguardian.com/environment/2013/nov/21/mass-walk-out-un-climate-talks-warsaw (last accessed June 2019).

a low-carbon economy, but also adopted an ambitious and historic resolution which states that

> Congress is committed to promoting an integrated approach to sustainable development through a just transition where social progress, environmental protection and economic needs are brought into a framework of democratic governance, where labour and other human rights are respected and gender equality achieved. (ITUC, 2010)

Climate change was high on the agenda of the ITUC's 3rd World Congress in Berlin in 2014, where Sharan Burrow was standing for re-election. The main Congress session on climate kicked off with an impassionate statement from the Fiji Trades Union Congress that focused on their future and their hunger for social justice. They also pointed out that these could not be achieved unless massive changes happened elsewhere, and exhorted union comrades from all over the world to do their best to make those changes happen. Intense debates followed on the 'action points' which would complement the four-year plan adopted by the Congress and presented by the General Secretary. The few voices that expressed concerns over risks for jobs were countered by other union leaders who called for ambition, for the sake of their children's future and jobs, and for an honourable place in history for the labour movement. As the Congress report states:

> The ITUC supports the moral imperative to both preserve an inhabitable planet and to profit from the jobs that climate action can deliver. We demand a commitment to a 'just transition' based on social dialogue from the workplace to the national level, with green skills and social protection guaranteed. To that end we will work to see an ILO standard to guide government and employer action. Equally, we will work to ensure that our own workers' capital is increasingly invested in the real economy, including in both industrial transformation and new green jobs. All jobs must be cleaner if we are to green our economy, and decent work must be at the heart of this transformation. As the world prepares for another deadline for a global agreement on climate by December 2015, the ITUC will mobilise our members and their communities to demand an ambitious and binding accord. Holding governments to account for climate is a top priority. (ITUC, 2014:9f)

The policies adopted at the Berlin Congress drove the last leg of union advocacy efforts in the UNFCCC on the road to the Paris Climate Conference (COP21). The ITUC's positions in climate negotiations were no longer those of a 'green fringe' but were now the official position for the entire movement, endorsed by all its elected representatives. There was now a clear mandate for the ITUC Secretariat to get affiliates to hold this line at the national level.

In the lead-up to the Paris Conference, various unions proceeded to transfer the key elements of the ITUC just transition strategy – as defined in the Congress resolution and in COP statements – and to advocate for their inclusion in national climate policies. In some countries, such as Argentina or Ghana, unions came together to collectively develop national just transition plans. Elsewhere, unions focused their efforts on existing climate policies and advocated for just transition measures to prevent any negative social impacts. This was the case in France where the Confédération Générale du Travail (CGT) and Confédération Française Démocratique du Travail (CFDT) influenced government plans for a heavy-duty vehicles tax (that, unfortunately, would later be scrapped by the government).

Various unions also developed sector-specific proposals, especially in the energy sector. In response to the government's decision to close 23 mining operations and suspend five others, Filipino energy unions, for instance, came up with a joint just transition plan for mining industry workers. The plan included measures to strengthen the social dialogue mechanism, build the capacity of institutions to implement new laws, ensure proper compensation measures, promote economic diversification, and guarantee social protection and insurance. All proposed measures draw on the core policy recommendations contained in the just transition concept note developed by the international labour movement and later endorsed by the ILO (ILO & SIDA, 2017:49).

It is worth mentioning that union-led climate conversations and proposals in many countries, notably in the Global South, went beyond the protection of the workers currently employed in the fossil-related extractive sector and stressed the need to invest in alternatives. They also highlighted the need to address major fault lines in the national economies, such as informal and precarious work, attacks against indigenous peoples' rights, or matters of access to essential services and their public provision.

PARIS AND BEYOND

The reference to just transition in the Paris Climate Agreement, and the fact that, four years after Paris, many stakeholders and observers continue to refer to the concept and to trade unions as key elements of any low-carbon strategy is not a coincidence but the culmination of over 15 years of intense efforts. These have involved both using the UNFCCC process to raise union awareness on the connections between climate and labour, and to simultaneously get other stakeholders in the climate debate to recognise that there can be no climate ambition without social justice.

Through the development of the concept of just transition, and the idea that it was inseparable from climate ambition, the trade union movement positioned itself as an anchor and engine for justice and equity in and beyond the workplace. Through the climate question, it was, in a sense, reconnecting with its roots. Historically, unions have always acted in the interests of society, and not just their members. In many ways, the climate struggle is also a struggle for justice. It is a struggle to defend those who are the least responsible for the climate crisis and yet have the most to lose. This includes but is not restricted to workers. As unions, we could not stay on the margins or turn a blind eye to the rest of humanity. We had to be on the right side of history.

As we have seen, a key to our success was our ability to strike the right balance between, on the one hand, raising the level of trade union ambition through tough and complex internal conversations, and on the other, engaging with others and getting them to engage with the social aspects of the low-carbon transition. And that balance has indeed been difficult to maintain following Paris.

Through its inclusion in the Paris Agreement and appropriation by a wide range of actors, just transition also became a less 'ideological' and less 'charged' concept. It would have been naïve to imagine that now that the concept had been legitimised through the UN, other voices would not attempt to appropriate it and promote a more 'sanitised' version of it. After Paris, a range of new understandings of just transition surfaced which, in certain cases, erased its links to the union movement. Our decision following Paris to develop partnerships with business networks and organisations, for instance, inadvertently contributed to the co-optation of just transition by less socially inclined interests. And we did this without pursuing our own internal engagement efforts, since, following Sustainlabour's closure in 2016, no other organisation was willing to 'invest' the time and energy to educate and empower unions, and strengthen their role in environmental debates. We had, in effect, lost touch with the fact that our ultimate goal was never to get just transition into the Paris Agreement. Just transition was not an end in itself but a means to an end. It was a vital tool to get unions and ultimately workers to be in a position to support ambitious climate action, and in the process, re-place the values of international solidarity and social justice at the heart of both the union and climate agendas.

REFERENCES

AFL-CIO (American Federation of Labor and Congress of Industrial Organizations). 2017. *Convention Resolution 55: Climate Change, Energy and Union Jobs*. 24 October. https://aflcio.org/resolutions/resolution-55-climate-change-energy-and-union-jobs (last accessed April 2019).

American Clean Energy and Security Act of 2009, H.R. 2454, 111th Congress, 7 July 2009. www.congress.gov/bill/111th-congress/house-bill/2454/text (last accessed April 2019).

Cassegard, C & H Thorn. 2017. 'Climate justice, equity and movement mobilization', in H Thorn, C Cassegard, L Soneryd & A Wettergren (eds), *Climate Action in a Globalizing World: Comparative Perspectives on Environmental Movements in the Global North*, pp.33–56. New York & London: Routledge.

Duyck, S, S Jodoin & A Johl. 2018. *Routledge Handbook of Human Rights and Climate Governance*, New York: Routledge.

Enkvist, PA, T Nauclér & J Rosander. 2007. 'A cost curve for greenhouse gas reduction', *McKinsey Quarterly*, February 2007. www.mckinsey.com/business-functions/sustainability/our-insights/a-cost-curve-for-greenhouse-gas-reduction (last accessed April 2019).

Foster, D. 2010. 'Blue-green alliance: Building a coalition for a green future in the United States', *International Journal of Labour Research*, vol. 2, no. 2, pp. 233–244. www.ilo.org/wcmsp5/groups/public/---ed_dialogue/---actrav/documents/publication/wcms_153352.pdf (last accessed April 2019).

ICFTU (International Confederation of Free Trade Unions). 2004. *Final Resolution. Globalisation, Decent Work and Sustainable Development*. ICFTU Eighteenth World Congress, Miyazaki, 5–10 December 2014. www.ilo.org/wcmsp5/groups/public/---dgreports/---wcsdg/documents/event/wcms_080296.pdf (last accessed April 2019).

ILO (International Labour Organization). 2015. *Guidelines for a Just Transition Towards Environmentally Sustainable Economies and Societies for All*. Geneva: ILO.

——. 2012. *Working towards Sustainable Development: Opportunities for Decent Work and Social Inclusion in a Green Economy*. Geneva: ILO. www.ilo.org/wcmsp5/groups/public/---dgreports/---dcomm/---publ/documents/publication/wcms_181836.pdf (last accessed April 2019).

ILO (International Labour Organization) & SIDA (Swedish International Development Cooperation Agency). 2017. *SIDA-ILO Partnership Programme*. www.ilo.org/wcmsp5/groups/public/---dgreports/---exrel/documents/genericdocument/wcms_369803.pdf (last accessed May 2019).

IPCC (Intergovernmental Panel on Climate Change). 2007. *Fourth Assessment Report*. www.ipcc.ch/assessment-report/ar4/ (last accessed May 2019).

ITUC (International Trade Union Confederation). 2014. *Building Workers' Power. Congress Statement*. 3rd ITUC World Congress, Berlin, 18–23 May 2014. https://congress2014.ituc-csi.org/3co-e-5-building-workers-power (last accessed April 2019).

——. 2010. *Resolution on Combating Climate Change through Sustainable Development and Just Transition.* ITUC 2nd World Congress, Vancouver, 21–25 June 2010. www.ituc-csi.org/IMG/pdf/2CO_10_Sustainable_development_and_Climate_Change_03-10-2.pdf (last accessed April 2019).

——. 2006. *Constitution.* Adopted at the Founding Congress, Vienna, November 2006. www.ituc-csi.org/IMG/pdf/Const-ENG-W.pdf (last accessed April 2019).

——. Trade Union Confederation of the Americas & Sustainlabour. 2012. *Trade Union Resolution on Labour and Environment.* 2nd Trade Union Assembly on Labour and the Environment. www.ituc-csi.org/IMG/pdf/resolution_-_2nd_trade_union_assembly_on_labour_and_environment.pdf.pdf (last accessed April 2019).

Martin Murillo, L. 2013. 'From sustainable development to a green and fair economy: Making the environment a trade union issue', in N Räthzell & D Uzzell (eds), *Trade Unions in the Green Economy: Working for the Environment,* pp. 29–40. London: Routledge.

NBC News. 2005. 'Bush puts jobs ahead of Kyoto', 15 February. www.nbcnews.com/id/6976284/print/1/displaymode/1098/ (last accessed May 2019).

Regulation (EC) No 1907/2006 of the European Parliament and of the Council of 18 December 2006 concerning the Registration, Evaluation, Authorisation and Restriction of Chemicals (REACH), establishing a European Chemicals Agency. https://eur-lex.europa.eu/eli/reg/2006/1907/2014-04-10 (last accessed April 2019).

Renner, M, S Sweeney & J Kubit. 2008. *Green Jobs: Towards Decent Work in a Sustainable, Low-Carbon World.* Nairobi: UNEP/ILO/IOE/ITUC. www.ilo.org/wcmsp5/groups/public/@dgreports/@dcomm/documents/publication/wcms_098504.pdf (last accessed April 2019).

Rosemberg, A. 2017. *Strengthening Just Transition Policies in International Climate Governance. Policy Analysis Brief.* Stanley Foundation, www.stanleyfoundation.org/publications/pab/RosembergPABStrengtheningJust-Transition417.pdf (Accessed 04 June 2019).

UNEP (United Nations Environment Programme). 2007. *Labour and the Environment: A Natural Synergy.* wedocs.unep.org/bitstream/handle/20.500.11822/7448/-Labour%20and%20the%20Environment_%20A%20Natural%20Synergy-2007739.pdf?sequence=3&isAllowed=y (last accessed April 2019).

Vidal, J. 2009. 'Copenhagen climate summit in disarray after "Danish text" leak', *The Guardian,* 8 December. www.theguardian.com/environment/2009/dec/08/copenhagen-climate-summit-disarray-danish-text (last accessed April 2019).

WEDO (Women's Environment and Development Organization). 2015. *COP21: Human Rights and Gender Equality.* https://wedo.org/cop21-human-rights-gender-equality/ (last accessed April 2019).

WWF (World Wide Fund for Nature). 2012. 'Doha climate talks bury international climate action in the desert', WWF, 13 December. wwf.panda.org/wwf_news/?207067/Doha-Climate-Talks-Bury-International-Climate-Action-in-the-Desert (last accessed April 2019).

2

Business in just transition: The never-ending story of corporate sustainability

Nils Moussu

INTRODUCTION

Transnational corporations and their representative associations are late participants in the debates surrounding just transition. Clearly not a driving force in its promotion, business had to respond to the successful lobbying efforts made by global union federations, such as the International Trade Union Confederation (ITUC), efforts that led to the establishment of just transition as an inevitable issue in transnational climate governance. In 2015, when the Preamble of the Paris Agreement took into account 'the imperative of a just transition of the workforce and the creation of decent work and quality jobs' (UNFCCC, 2015), just transition was almost entirely missing in the positioning and proposals from business (see We Mean Business, 2015a). Yet, prominent corporate groupings have since been able to catch up and develop their own understanding of just transition, ensuring their readiness for the 2018 'Just Transition COP (Conference of the Parties)' in Poland (COP24, Katowice). The growing popularity of just transition among business remains a new and evolving trend. It represents nevertheless a notable and important evolution of what will be described in this chapter as the 'business in transition' narrative – a story in which workers had previously often been completely absent.

From a critical standpoint, this evolution still remains challenging to grasp fully. On the one hand, transnational corporations obviously have a key role to play in the transition from a fossil-based economy to a low-carbon or decarbonised world. As major employers, either directly or through their global production networks, and especially in carbon-intensive economic sectors, corporations make daily technological, political and investment choices that facilitate or hinder this transition. Considering this critical position, as Snell (2018:554) puts it, 'The role and responsibility of private sector actors ... is surprisingly absent from much of the current debate about [just transition].' Going beyond a narrow focus on state regulations and environmental or labour market policies is all the more important as business managed to be considered the 'co-governors' and leaders of the international climate change (Paris Agreement) and sustainability (Sustainable Development Goals) processes in an unprecedented way (Benabou et al., 2017; Chan et al., 2019). On the other hand, one should wonder if just transition is compatible at all with the very nature of transnational corporations – which extract, exploit and expand in order to generate more sales and profits (Dauvergne, 2018) – even in just transition's lighter 'status quo' and 'managerial reform' approaches (JTRC, 2018). In short, how much faith can be placed in transnational corporations to lead sustainability and just low-carbon transition through their usual means (self-regula-tion, market mechanisms, private certification, sustainability reporting, etc.)? And is this emerging 'corporate just transition' approach simply diluting, perverting, or co-opting more authentic versions that strive to contribute to social and ecological emancipation?

To start answering these questions, this chapter retraces the embedding of a just transition discourse into the prior 'business in transition' story – as it was narrated before the recent globalisation and proliferation of the just transition idea. It does so by focusing on the role of large trans-national business associations and coalitions representing corporate members from multiple countries and sectors in climate negotiations. Unlike narrower corporatist and mono-sectoral associations, these large business groupings are conceived in a Gramscian-inspired way as collec-tive intellectuals, organically linked with transnational capitalist social forces, and essentially performing a doctrinal work (Carroll & Carson, 2003). While representing corporations, these business groupings are not to be confused with their corporate members. As 'business ethicists' (Abend, 2014), they offer a 'discursive service' to their affiliates on which they depend financially. This service must be simultaneously self-critical enough of the current environmental record of corporations, framed in a business-friendly way (e.g. risks and opportunities), as well as bold and ambitious for the (future) corporate environmentalism, described as able to literally save the world. From these actors, the embedding of some forms of just transition discourse into corporate environmentalism can be understood as a necessary evolution to sustain their broad narrative of 'business in transition' by covering all the possible issues in climate debates; as these issues could otherwise create easy avenues for criticis-ing transnational corporations or capitalism in general. At the same time, this chapter will characterise this process as a relatively painless evolution that simply adds or specifies one aspiration to a 'win-win' story, which is already built to include a wide range of objectives (people, planet, profit) in a non-conflicting way. It will do so by, first, explaining the narrative of 'business in transition' and discussing its main features. Then, it will retrace the adoption of a just transition discourse by prominent business groupings in the run-up to and during COP24. Finally, it will assess the extent to which the idea of just transition has transformed the 'business in transition' narrative so far.

THE 'BUSINESS IN TRANSITION' NARRATIVE: BEFORE JUST TRANSITION

The past decades have seen transnational corporations, their executives and representative organisations devoting significant efforts to be recognised as sustainability leaders (Wright & Nyberg, 2015; Dauvergne, 2018). In climate negotiations, corporations have gradually built a progressive and large 'business voice', far from the defensive and obstructive stance still dominant in the 1990s (Levy & Egan, 2003). This mobilisation reached its climax at the Paris Climate Conference (COP21), with the strong involvement of corporate representatives and their associations, as well as multiple business pledges and commitments to take action through numerous private and hybrid governance initiatives (Benabou et al., 2017). The actual contribution of business actors to bridge the emissions gap to ensure that global warming stays well below 2°C remains highly uncertain and difficult to assess (Hsu et al., 2019). And business's status of 'climate hero' is strongly contested by civil society organisations, which are campaigning against the influence of big polluters from the fossil fuel industry within climate negotiations.[1] Nonetheless, corporations are now widely seen as a 'force of good'[2] that 'must lead the way on climate action' (Fabius & Figueres, 2015) given the cumbersome political negotiation process between states. Even more ambitious, business presents itself as having 'to take the lead in saving the world' (Bakker, 2013).

Such bold claims could appear as totally irrelevant or even outrageous, given the current environmental records of transnational corporations, their heavy responsibility in global warming (Heede, 2014), and the collective failure in attempts to seriously cut greenhouse gas emissions. Yet,

---

[1]   See, for instance http://kickbigpollutersout.org/about/ (last accessed January 2019).

[2]   As expressed by Lise Kingo, CEO and Executive Director, UN Global Compact, during an International Chamber of Commerce press conference organised at the COP21, 8 December 2015.

those claims and the legitimacy they gained in the transnational climate governance space have to be understood as part of the long history of corporate and liberal environmentalism (Bernstein, 2001). Patiently and systematically built by corporate representatives and their organic intellectuals, the corporate environmentalism movement evolved around key concepts, such as sustainable development, eco-efficiency, corporate social responsibility, green economy, etc. But beyond the fads and fashion of these notions, corporate environmentalism is based on a narrative told repeatedly by a growing fraction of business since the 1990s: the story of 'business in transition' (see Moussu, 2017 for an in-depth analysis of this narrative). While the idea of narrative is often conflated with other concepts such as discourses, frames, or perspectives, I refer next to narrative in two specific ways.

First, following recent work emphasising the importance of time in international relations, narratives are understood as the discursive outcome of the activity of 'timing'. Hom (2018:70) defines timing as 'the practical efforts by which social agents establish meaningful relationships between processes of change so that they unfold in ways conducive to orientation, direction, and control'. In climate debates, while texts and talks come in several genres (reports, position papers, newspaper articles, etc.), the numerous scenarios about the transition (/transformation/ race/pathways/journey ...) towards a green (/low-carbon/sustainable/ net-zero emissions ...) economy clearly allows for establishing such relationships between processes of change. Crucial here is the idea of control, since the 'business in transition' narrative is used by corporate representatives as a discursive vehicle pursuing at least three political objectives: 1) to legitimise corporate activities and safeguard business's social licence to operate in the face of climate change; 2) to increase corporate power on problem definition within climate governance, and 3) to define the economic or technological solutions, and the (non)regulations negotiated to tackle this issue. Hence, this narrative is not a mere lobbying tool used to meet the narrow corporatist needs of corporations. Rather, it is used to enter the struggle for hegemony, and is developed by organic and collective intellectuals who need, in the words of Gramsci, to organise 'the general system of relationships external to the business itself' (Gramsci, 1971:135). This task is performed by an entrepreneurial elite in order to 'bring about not only a unison of economic and political aims, but also intellectual and moral unity, posing all the questions around which the struggle rages not on a corporate but on a universal plane' (Gramsci, 1971:406). In this regard, if the tensions between accumulation and legitimation have always existed (Paterson, 2010), climate change creates new avenues for criticising business corporations or capitalism in general (Chiapello, 2013). It also creates new opportunities for corporate actors to seize ecological criticisms and present themselves as the main architects of a 'safe and prosperous future' (We Mean Business, 2015b).

Second, from a text-linguistic perspective, narratives are prototypical text sequences that offer a storyline or a plot consisting of temporally ordered and causally related events (Adam, 2011). They are usually made of five components or subsequent moments (Fløttum & Gjerstad, 2017): first, the initial situation, before imminent action; second, complications, interrupting the initial situation and setting the narrative process in motion; third, possible reactions and likely courses of action in response to the complications, and finally, resolution and final situation, after the recently completed process. The point here is not to claim that all discourses dealing with climate change display the classical narrative structure, as most fairy tales or novels do. However, many of these discourses feature a distinctive set of temporal and narrative turns that make them eligible for a narrative analysis. In climate change narratives, global warming itself is typically the complication factor of the story, but it can in turn create new complications or worsen typical complications faced by characters. Importantly, social issues have always been part of the 'business in transition' narrative since its first versions (e.g. Schmidheiny, 1992), which relied heavily on the common definition of sustainable development and its environmental, social and economic components. Issues ranging from human rights, growing inequalities, and access to healthcare, clean water, sanitation, energy and education were often included, together with growing concerns about the loss of popular support and confidence in transnational corporations, the market economy and capitalism. However, these complications, or risks in the words of business, were not addressing the issue of social justice at the company level with regards to the transition towards a decarbonised world. The business story had to wait until the globalisation and proliferation of the just transition idea in the 2010s to integrate this specific issue (see the next section).

Yet, what are the main components of the 'business in transition' narrative? When reading the corporate environmentalism literature, one can only be stunned by its simplicity and stability over time. It is basically made up of four related moments. First, a 'business as usual' and unsustainable *initial situation* is presented. This situation is then disturbed by alarming environmental, social and economic *complications* (risks, trends ...) that create a strong imperative for urgent action. Next, while business is said to have already begun to *react to these complications*, new courses of action are suggested in order to do more and to scale up. Finally, and given the appropriate regulatory framework established (i.e., the free market), business is presented as willing to take the lead in building a more sustainable world 'in which the global population is not just living on the planet, but living well and within the limits of the planet' (WBCSD, 2010: Executive Summary) – *resolution and final situation*.

This narrative makes an extensive use of a specific lexicon related to roads and journeys (e.g. course, path, road, pathway, race, destination, direction ...), which Milne and colleagues (2006) call the 'journey metaphor'. The use of this lexicon allows business to be introduced in the story as an active agent of change, in motion, and well on the road towards sustainability. As the main character of the narrative, unsurprisingly, business describes itself as a 'frustrated hero' willing to do more, provided that the 'right regulations' are put in place (e.g. regulations that favour price signals through carbon pricing over command-and-control policies). Through this self-description, business is able to escape the role of the villain and to promote the view that climate talks should not be *about* business, but conducted *in dialogue with* business, and that these talks should produce regulation *for* business instead of regulation *of* business. Overall, the takeaway message is that business has changed, is changing, and will keep changing even more. Obviously, this narrative tells 'what should be the case, not what is the case' (Boltanski & Chiapello, 2005:58), and the internal and discursive temporality of the narration should be compared with actual changes on the ground. Nevertheless, to remain plausible, this narration cannot afford to be just a fictional discourse blind to its political environment. An example of such need is the adoption of a just transition discourse by prominent business associations discussed in the next section. Of course, what is discussed here from a discourse-analysis perspective is accompanied by processes of change at different levels of a hegemonic historical structure – i.e., ideas and discourses have to fit to a certain extent social, institutional and material configurations of forces (Cox, 1981). For instance, building a plausible business narrative on just transition also involves forging alliances and co-opting other social forces, even at the cost of corporatist sacrifices. In the words of Gramsci:

Undoubtedly ... the leading group should make sacrifices of an economic-corporate kind [in order to establish hegemony]. But there is also no doubt that such sacrifices ... cannot touch the essential; for though hegemony is ethical-political, it must also be economic, must necessarily be based on the decisive function exercised by the leading group in the decisive nucleus of economic activity. (Gramsci, 1971:373)

If what will be described as a painless adoption of the idea of just transition in the 'business in transition' narrative could be explained by the level of opposition business is facing in transnational climate governance, the remainder of this chapter will stay at the level of discourse analysis.

LOOKING FOR A JUST TRANSITION BUSINESS CASE

Three years after the adoption of the Paris Agreement, the idea of just transition reached the status of a legitimate and mainstream issue in international climate negotiations (JTRC, 2018). Reflecting this process, the use of a just transition discourse, still primarily associated with the union movement, was adopted by a wide array of actors, including business groupings in their position papers released in the run-up to and during the COP24 (see We Mean Business, 2015a). Essentially directed towards policy makers, this adoption can be interpreted as a mere strategic move to avoid being excluded from the just transition discussion. Nonetheless, it also shows how specific business groupings have performed a substantive groundwork on the implementation of just transition at the company level. After providing an overview of business recommendations regarding just transition within the United Nations climate regime, this section discusses the main features of what is promoted as the 'business case' for just transition.

When entering the international climate negotiations, new issues open new social, institutional and discursive spaces: side events gathering diverse actors are dedicated to these topics, specialised fora and processes are often created within official negotiations (e.g. the Improved Forum on the Impact of the Implementation of Response Measures), and new terms or discourses are released and framed. While complex and uneven, this process of 'climatisation' (Aykut et al., 2017), when successful, creates an absolute need to fill these spaces for actors who strive to cover and influence the main issues on the agenda. When the idea of just transition found its way into negotiation documents, large business groupings simply could not afford to ignore it. The result is easily observable in the business statements released for the 2018 climate negotiations in Poland (COP24, Katowice).

Take for instance an individual association such as the World Business Council for Sustainable Development (WBCSD), a major grouping representing about two hundred large corporations. This association considers just transition as one of the 'enablers' – together with carbon pricing, and ambitious adaptation and resilience efforts – allowing for the limitation of 'global warming to well below 2°C, with 1.5°C as the new North Star' (WBCSD, 2018). Directed towards policy makers,

WBCSD's recommendations state that 'we need to ensure a just transition so that no one is left behind, to create jobs that are green and decent and to ensure that communities are thriving and resilient' (WBCSD, 2018). Referring to the idea of decent work and the imperative of leaving no one behind, this example can be characterised as the minimal and standard statement of progressive business groupings.

Interestingly, a traditionally more conservative association, the International Chamber of Commerce (ICC), produced three key 'Principles for a Just Transition' (ICC, 2018). The first one advocates for the inclusion of just transition of the workforce into the Nationally Determined Contributions (or NDCs, which are countries' planned climate actions to be reported under the Paris Agreement). This first principle is shared by global unions such as ITUC (2018) and mentioned in the 'Solidarity and Just Transition Silesia Declaration' developed by the Polish COP24 Presidency.[3] The second principle is a general public policy recommendation, requiring policy makers to 'enable workers and their communities to have the skills capabilities and investments needed to thrive in the face of transformative change towards a sustainable and inclusive economy' (ICC, 2018). Broadly in line with the *Guidelines for a Just Transition* developed by the International Labour Organization (ILO, 2015), this second principle details various labour market programmes (e.g. training programmes, income support, social insurance programmes, etc.) and complementary policies (e.g. education, housing, etc.). Finally, the third principle expresses business's strong and recurrent demand to be included 'as key stakeholder and agent for the delivery of a just transition' in 'climate policy planning at all levels' (ICC, 2018). Typical of demands included in this kind of position paper, this principle promotes the unique position of business – 'aware of its social responsibilities' and having the adequate skills to help and advise governments. While this last principle can be related to social dialogue and tripartite negotiations between governments, unions and employers (business and 'other stakeholders' in the document), it also shows an underlying willingness to regain control on the just transition debates in climate negotiations.

This demand is even more obvious in the position paper developed by the United States Council for International Business, and the Major Economies Business Forum on Energy Security and Climate Change – this last association representing the International Organisation of Employers among other national and sectoral associations. In their

---

3 See https://cop24.gov.pl/presidency/initiatives/just-transition-declaration/ (last accessed January 2019).

paper, they recommend the establishment of a new business platform within the United Nations Framework Convention on Climate Change (UNFCCC) (USCIB & BizMEF, 2018). Focused on 'Just Transition to Transformation', and led by and for business, this platform would 'include and go beyond [considerations] identified in "just transition" involving workers, to encompass the various roles of the private sector to accelerate transformational change' (USCIB & BizMEF, 2018:5). In this case, the use of just transition is almost reduced to a keyword, arguably mobilised to obtain an old request from business, i.e. having a long-lasting dedicated consultative platform within the UNFCCC and linked with its Secretariat (beyond existing avenues such as the Non-state Actor Zone for Climate Action, and the UNFCCC Talanoa dialogue concluded at the COP24, for instance).

From minimal to more developed uses of just transition, these examples illustrate how business groupings have adapted to this new idea and mobilise it now in international climate negotiations. References to just transition are now fairly common in individual multisectoral business associations' statements (e.g. WBCSD, ICC), as well as in statements of business coalitions. This last case can be exemplified by the We Mean Business coalition 'Calls for Increased Climate Ambition' (We Mean Business, 2018). As one of the major business voices in climate negotiations, We Mean Business unites many partner organisations, including the WBCSD, Business for Social Responsibility (BSR), CDP (formerly the Carbon Disclosure Project), the Coalition for Environmentally Responsible Economies (Ceres), The B Team, The Climate Group, and The Prince of Wales's Corporate Leaders Group. Launched in 2014, funded by the IKEA Foundation, and claiming to work with more than six million corporations worldwide, We Mean Business plays a central role in aligning business and creating a strong 'business voice' within the UNFCCC. In its vision for 'resilient, just, zero-carbon economies', We Mean Business presents a detailed set of requirements for policy makers regarding national policies (employment, education, social protection), specific regional and sectoral policies (e.g. in high-emitting sectors or green industries), and tripartite social dialogue. These requirements clearly stem from the work carried out by two of We Mean Business partner organisations, namely The B Team and BSR, a work that deserves a detailed discussion.

The B Team describes itself as 'a global group of business leaders' developing a Plan B to transform business into 'a driving force for social, environmental and economic benefit', as opposed to a Plan A 'where companies have been driven by the profit motive alone'.[4] Co-founded by Richard Branson (Virgin Group) and Jochen Zeitz (former CEO of Puma), The B Team is financed by various philanthropic foundations (Ford, Rockefeller, Tiffany and Co., Virgin Unite), corporations (Unilever, Kering Group), and individuals (e.g. Guilherme Leal, Strive Masiyiwa). This group gathers 'leaders', from transnational corporations of course, but also from various backgrounds. The first group includes for example The B Team's Chair, Paul Polman (also Chair of the ICC, Vice-chair of the UN Global Compact Board, commissioner of the Business & Sustainable Development Commission, former CEO of Unilever and Chairman of the WBCSD). This group also includes Oliver Bäte (CEO, Allianz Group), Emmanuel Faber (Chairman and CEO, Danone), Isabelle Kocher (CEO, Engie), and François-Henri Pinault (Chairman and CEO, Kering). They are accompanied by individuals with diverse profiles, such as The B Team's Co-chair Sharan Burrow (General Secretary, ITUC) and other prominent figures such as Christiana Figueres (Mission 2020, former UNFCCC Executive Secretary), Mary Robinson (President, Mary Robinson Foundation— Climate Justice), Gro Harlem Brundtland (Deputy chair, The Elders), and Yolanda Kakabadse (former President of the World Conservation Union and of the WWF International).

This diversified membership almost raises the question of the very 'business' nature of The B Team. In any case, it allows this organisation to create strong links with the work carried out by non-business actors. A case in point is the report 'Just Transition: A Business Guide' published jointly by the Just Transition Centre (JTC) and The B Team (2018), which certainly benefited from the pivotal position of Sharan Burrow, ITUC General Secretary and B Team leader. Established in 2016 by the ITUC and the European Trade Union Confederation, the JTC was designed from the beginning to bring together workers and their unions, businesses, governments, civil society and communities, as shown in its initial 'Call for Dialogue' endorsed, inter alia, by The B Team and We Mean Business (ITUC, 2016). Materialised in the joint JTC & B Team report, this collaborative stance has enabled the ITUC to impose its own definition of just transition, taken word for word (see ITUC, 2017:6) by a major business group in the very first attempt to tackle this issue:

> For companies, a just transition is an enterprise-wide process to plan and implement companies' emissions reductions efforts, based on social dialogue between workers and employers. This includes a

---

4   See www.bteam.org/about/ (last accessed January 2019).

company's supply chains. The goal is to reduce emissions and increase resource productivity in a way that retains and improves employment, maximizes positive effects for workers and communities, and allows the company to grasp the commercial opportunities of the low-carbon transition. (JTC & The B Team, 2018:2)

The above definition, mobilised since then as a model for further elaboration of 'corporate just transition', is consistent with what has been identified as the ideal-typical 'managerial reform' approach to just transition (JTRC, 2018). This approach tries to articulate and combine three main elements: the limitation of global warming to well below 2°C above pre-industrial levels (planet); social fairness and justice in the transition towards a decarbonised or net-zero world (people), and all this within the existing global political economy (profit). Very similar definitions and approaches were adopted in the BSR report 'Climate and the Just Transition, The Business Case for Action' (Wei, 2018) released during COP24.[5] In developing this approach, both The B Team and BSR report strive to show that the purpose of corporations, in their enlightened self-interest, is not solely the profit motive – in short, 'the business of business is more than business' (WBCSD, 1997:27). The main messages of these two JTC & B Team and BSR reports are now discussed together.

The demonstration of the 'business case for just transition' in these two reports relies on recurring arguments of the familiar 'business in transition' narrative. First, a brief overview of possible external pressures on corporations is given. Just transition is introduced as 'a key requirement of the Paris Agreement' (JTC & The B Team, 2018:2) and linked with multiple international frameworks, guidelines and principles developed by international organisations such as the ILO or the United Nations (see ILO, 2015; OECD, 2011; UN, 2015; UNOHCHR, 2011). More importantly, the pressure coming from the investors and the growing imperative of 'full disclosure' is emphasised (see Robins et al., 2018). For corporations, disclosing non-financial information on environmental, social and governance issues is said to be 'increasingly important for investors and rating agencies' (JTC & The B Team, 2018:6).[6] To be disclosed in this

---

5 This report was funded by the IKEA Foundation and We Mean Business (Wei, 2018). BSR is a We Mean Business partner organisation that gathers more than 250 transnational corporations in a conventional membership-based business association. See www.bsr.org/en/about (last accessed January 2019).

6 On non-financial disclosure, the main initiative mentioned are the CDP, the Global Reporting Initiative, the Sustainability Accounting Standards Board, and the Workforce Disclosure Initiative, aligning their requirements with the Task Force on Climate-related Financial Disclosures recommendations (TCFD, 2017).

regard are the 'risks of transition' for business, workers and communities, including 'transparent information about potential economic and employment dislocation and plans to address this' (Wei, 2018:16).

Second, the transition risks and opportunities are detailed, forming the bulk of the 'business case for just transition' and following the narrative structure in which complications are followed by reactions and solutions. The line of argument is straightforward: 'individual companies that contribute to a just transition will better manage the risks from a transition to the low-carbon economy and capitalize on related opportunities' (Wei, 2018:7). Adapted from the recommendations developed by the Task Force for Climate-Related Financial Disclosure (TCFD, 2017), four main risks are distinguished (JTC & The B Team, 2018:5–6; Wei, 2018:14–15):

1) Policy and legal risk, which concerns law violation and related legal actions, as well as misalignment with new climate regulations (e.g. future increase in carbon price);
2) Technology risk, which concerns high cost of retraining or hiring schemes due to unplanned technological shifts, as opposed to planned schemes developed with workers and their unions and benefitting from their knowledge and insights;
3) Market risk, which concerns increased costs and threatened revenues and jobs due to poor labour practices and social dialogue that affect corporations' ability to adapt to quick market shifts, and
4) Reputation risk, which concerns negative impacts on workers' recruitment and retention, as well as on brand and customer perception due to poor labour practices.

These risks are associated with costs that can be avoided if well managed. But, also showcased in the two reports, new economic opportunities are opened when investing in just transition. Based on short 'case studies' (Enel, Siemens, AGL Energy, PG&E among others), these opportunities depart from any idea of social justice and equity to highlight traditional business concerns: increased 'employee productivity, creativity and flexibility', facilitated 'adjustments in wages and working time', and improved 'customer loyalty and brand recognition' (Wei, 2018:15). But beyond such direct economic opportunities, broader benefits are expected for corporations. Investing in just transition is said to improve their social licence to operate, and to allow business to help 'shape regulation and legal reforms' with governments and unions (JTC & The B Team, 2018:5).

Finally, both the JTC & B Team and BSR reports develop recommendations and principles to implement just transition at the company level and grasp the related economic opportunities. In the JTC & The B Team report (2018:7–9), these recommendations are divided in three parts – 'engage, plan, enact' and emphasise the need for long-term planning of emission reductions and their impacts on workers and communities, as well as tripartite negotiations and social dialogue. For instance, the establishment of 'joint management-labour environment committees' is encouraged. At the company or sector levels, these committees would agree on options to engage in pathways to net-zero emissions, evaluate the consequences of climate actions on workers, and develop training services including 'initial learning in green business practices, continuous learning in occupational safety and health and continuous reskilling and upskilling in environmentally friendly technology and innovation' (JTC & The B Team, 2018:9).

Reviewing these reports project a strong overall sense of action: business did the job! It is now fully equipped with a definition of just transition, a related business case, and principles as well as recommendations for its implementation. This new 'corporate just transition' even materialises in actions, as demonstrated by the Pledge for a Just Transition to Decent Jobs announced during the Global Climate Action Summit held in September 2018 in San Francisco (The B Team, 2018). Through this pledge, renewable energy producers and developers (Enel and Ørsted) commit to provide four standards to their employees, and renewable energy buyers (Autodesk, Safaricom and Unilever) commit to procure electricity from producers adopting these standards: 'Social dialogue with workers and their unions; fundamental rights, including the ILO core labour standards and ILO occupational health and safety standards; social protection, including pension and health; and wage guarantees including prevailing wage rates for skilled workers in the relevant industries' (The B Team, 2018). Yet, despite the growing popularity of the just transition discourse among business actors, the embedding of such a discourse into the 'business in transition' narrative, the ambiguities and long history of this narrative, and its weak performative power, suggest that the 'corporate just transition' should be analysed with great caution.

THE 'BUSINESS IN TRANSITION' NARRATIVE: AFTER JUST TRANSITION

As described above, when trying to control environmentalism and climate governance, transnational corporations and their representatives enter a discursive struggle and create powerful narratives. As stated in the BSR report on just transition, this process is about 'unify[ing] discourse' and 'creating a shared vocabulary' (Wei, 2018:3) that simultaneously sets the limits of the legitimate issues and permissible criticisms, while excluding more radical approaches, subaltern groups and alternative futures. In this respect, the embedding of a just transition discourse into the 'business in transition' narrative can be simultaneously characterised as a a notable evolution, and a painless and limited evolution.

More specifically, the idea of just transition does indeed expand the scope of corporate theories of change or transition stories. Previously, workers were described in a static and passive way as mere stakeholders for corporations in general, and as part of the broad human dimension of sustainable development. Now, the social dimensions of the low-carbon transition at the company and workplace levels also form part of the story; and issues related to social and tripartite negotiations, education and training, social protection, and occupational safety and health, are included. But the story goes beyond these issues and expands to communities as well as corporations' global production networks, even if those dimensions have yet to be developed in greater detail. Still, the inclusion of these new issues into the 'business in transition' narrative marks a notable evolution, since every step that contributes to extend its scope and to characterise the future of its main characters creates standards against which corporate behaviour can be assessed and criticised.

Yet, this evolution also represents an easy and painless move for business. Corporate environmentalism has always highlighted the employment gains that 'green growth' and a transition to a low-carbon economy could create (e.g. BSDC, 2017). The adoption of a managerial approach to just transition centred on jobs is thus not surprising for a 'corporate just transition' explicitly developed in order 'to counter inaccurate arguments that climate action results in economic vulnerability and job losses' (Wei, 2018:19). Hence, connecting to a union's discourse highlighting the compatibility between protecting the planet and protecting workers was not a too complicated task for business interests that have always used a win-win rhetoric (good for the planet, for the people, for business, for growth, etc.).

This win-win rhetoric is at the core of the 'business in transition' narrative and the business case for just transition. It appears mainly in discourses through a single formula (e.g. sustainable development) or a list of objectives loosely linked with each other. Take for instance the following statement: '[just transition] requires job creation through seizing new economic opportunities, while reducing the disruption people and communities face in the transition away from high-carbon business models' (Wei, 2018:7). This use of concessive clauses and specific wording (such as 'without compromising', 'while', 'combining', etc.) to stitch together potentially contradictory objectives into a single formula or sentence is typical of win-win discourses on sustainable development (Krieg-Plan-que, 2010). In this way, conflicts and trade-offs between objectives are hidden, and key political economy questions of 'who wins, who loses, how and why' (Newell & Mulvaney, 2013:133) are avoided. In the case of just transition, objectives are numerous: 'reduc[ing] greenhouse gas (GHG) emissions ... enabling economic vitality and ensuring adherence to global labor standards ... enhanc[ing] climate resilience for communities ... cultivat[ing] effective participation in the social dialogue' (Wei, 2018:3). But in this case, too, difficult questions such as who will finance and bear the costs of the transition towards a low-carbon economy (corporations, shareholders, workers, states ...) remain unanswered. In other words, the business case for just transition remains weak compared to the business case for business as usual and 'unjust transition'. Drawing on selective or anecdotal cases, as well as general figures that are not specific to an economic sector, a country, or a corporation, this business case for just transition is based, furthermore, on the belief in an often narrow and limited non-financial reporting to drive the transition.

Made up of 'truths, half-truths and illusions' (Dauvergne, 2016), the 'business in transition' narrative has the ability to incorporate new objectives relatively easily. This is because the narrative structure itself allows for keeping some of its key features blurred and ambiguous. For instance, the main character of the business narrative is never clearly defined, alternatively referred to as business, businesses, companies, many companies, smart companies, a growing number of leading companies, etc. Such a blurred definition of the main character of the narrative immediately raises a question: how many companies is 'many companies', and who is really included in the perimeter of 'business'? This underdefined notion of 'business' makes it easy to claim, for instance, that business has begun to engage with just transition and is now adopting the Pledge for a Just Transition to Decent Jobs (The B Team, 2018). This claim would be true and not true at the same time, while, in fact, these are a few business associations acting as organic and collective intellectuals that have developed a corporate approach to just transition adopted by a few corporations, and that are now taking a pledge whose actual translation into reality remains uncertain and hard to assess. A second ambiguity concerns the exact stage business has reached in the transition process. The business narrative often claims that business is moving but stays very cautious about giving any precise timeline. In this regard, while the 'journey metaphor' of sustainability or just transition generates a spatial dimension (going from point A to point B), this metaphor also creates a temporal dimension. This last dimension allows for the narrativisation of business discourse: on the road towards sustainability, one should first decide to leave, the journey can take more or less time, it can include multiple intermediate stages, the date of arrival is more or less near, etc. Nevertheless, in the business narrative, the plot and overall temporality of the story – between past, present and future events – are always vague, as well as the final destination or moment.

To conclude, the very nature of the 'business in transition' narrative explains its long-lasting success, and the ease with which this narrative can integrate new concepts and ideas such as just transition. Yet, for movements and social forces trying to mitigate corporate power, the inclusion of just transition in corporate environmentalism can serve as a critical entry point. 'Corporate just transition' can indeed be used to force business to clarify its story: who exactly is doing what (specific corporations in precise economic sectors, workers, states, etc.)? Who should be included in just transition (workers and their unions, people and communities, global production networks, women, etc.)? When are things happening, and where? And who finances, benefits and bears the costs of the transition? Moreover, understanding the long history of corporate environmentalism can help to understand 'corporate just transition' as simply the last embodiment of the never-ending 'business in transition' narrative. Indeed, connecting the internal temporality of this narrative with the temporality of actual changes in the face of climate change can lead to a more radical critique: transnational corporations have been able to incorporate new ideas such as just transition and many others before in their discourses, but their real business is first and foremost the pursuit of profits, growth and power. In this view, out of the fatigue of listening for decades to a similar and never-ending 'business in transition' story, the main task for antagonistic social forces would be to help recapture this narrative.

REFERENCES

Abend, G. 2014. *The Moral Background. An Inquiry into the History of Business Ethics*. Princeton, NJ & Oxford: Princeton University Press.

Adam, JM. 2011. *Les textes: Types et prototypes* (3rd edn). Paris: Armand Colin.

Aykut, S, J Foyer & E Morena (eds). 2017. *Globalising the Climate. COP21 and the Climatisation of Global Debates*. London & New York: Routledge.

Bakker, P. 2013. *The (R)Evolution of Capitalism*. Cambridge Distinguished Lecture, Institute for Sustainability Leadership (February 2013). Online video, viewed January 2019, www.youtube.com/watch?v=5lN8iLeqL-w.

Benabou, S, N Moussu & B Müller. 2017. 'The business voice at COP21: The quandaries of a global political ambition', in S Aykut, J Foyer & E Morena (eds), *Globalising the Climate. COP21 and the Climatisation of Global Debates*, pp. 57–74. London & New York: Routledge.

Bernstein, S. 2001. *The Compromise of Liberal Environmentalism*. New York: Columbia University Press.

Boltanski, L & E Chiapello. 2005. *The New Spirit of Capitalism*. Translated by G Elliott. London & New York: Verso.

BSDC (Business & Sustainable Development Commission). 2017. *Better Business, Better World*. London: BSDC. http://report.businesscommission.org/uploads/ BetterBiz-BetterWorld.pdf (last accessed April 2019).

Carroll, WK & C Carson. 2003. 'Forging a new hegemony? The role of transnational policy groups in the network and discourses of global corporate governance', *Journal of World-Systems Research*, vol. 9, no. 1, pp. 67–102.

Chan, S, I Boran, H van Asselt, G Iacobuta, N Niles, K Rietig, M Scobie, JS Bansard, D Delgado Pugley, LL Delina, F Eichhorn, P Ellinger, O Enechi, T Hale, L Hermwille, M Honegger, A Hurtado Epstein, S La Hoz Theuer, R Mizo, Y Sun, P Toussaint & G Wambugu. 2019. 'Promises and risks of nonstate action in climate and sustainability governance', *Wiley Interdisciplinary Reviews: Climate Change*, e572. https://doi.org/10.1002/wcc.572.

Chiapello, E. 2013. 'Capitalism and its criticisms', in P du Gay & G Morgan (eds), *New Spirits of Capitalism? Crises, Justifications, and Dynamics*, pp. 1–33. Oxford: Oxford University Press.

Cox, RW. 1981. 'Social forces, states and world orders: Beyond International Relations Theory'. *Millennium – Journal of International Studies*, vol. 10, no. 2, pp. 126–155.

Dauvergne, P. 2018. *Will Big Business Destroy Our Planet?* Cambridge & Medford: Polity Press.

——. 2016. 'The sustainability story: Exposing truths, half-truths, and illusions', in S Nicholson & S Jinnah (eds), *New Earth Politics. Essays from the Anthropocene*, pp. 387–404. Cambridge, MA: The MIT Press.

Fabius, L & C Figueres. 2015. 'Global businesses must lead the way on climate action', *The Guardian*, 19 May. www.theguardian.com/environment/2015/may/19/global-businesses-must-lead-the-way-on-climate-action (last accessed April 2019).

Fløttum, K & Ø Gjerstad. 2017. 'Narratives in climate change discourse', *Wiley Interdisciplinary Reviews: Climate Change*, vol. 8, no. 1, e429. https://doi.org/10.1002/wcc.429.

Gramsci, A. 1971. *Selections from the Prison Notebooks*. Edited and translated by Q Hoare & G Nowell Smith. New York: International Publishers.

Heede, R. 2014. 'Tracing anthropogenic carbon dioxide and methane emissions to fossil fuel and cement producers, 1854–2010', *Climatic Change*, vol. 122, no.1–2, pp. 229–141.

Hom, AR. 2018. 'Timing is everything: Toward a better understanding of time and international politics', *International Studies Quarterly*, vol. 62, no. 1, pp. 69–79.

Hsu, A, N Höhne, T Kuramochi, M Roelfsema, A Weinfurter, Y Xie, K Lütkehermöller, S Chan, J Corfee-Morlot, P Drost, P Faria, A Gardiner, DJ Gordon, T Hale, NE Hultman, J Moorhead, S Reuvers, J Setzer, N Singh, C Weber & O Widerberg. 2019. 'A research roadmap for quantifying non-state and subnational climate mitigation action', *Nature Climate Change*, vol. 9, no. 1, pp. 11–17.

ICC (International Chamber of Commerce). 2018. *ICC Principles for a Just Transition. Climate Change and the Economic and Social Viability of Communities.* Paris: ICC. https://iccwbo.org/publication/icc-principles-just-transition/ (last accessed April 2019).

ILO (International Labour Organization). 2015. *Guidelines for a Just Transition Towards Environmentally Sustainable Economies and Societies for All.* Geneva: ILO.

ITUC (International Trade Union Confederation). 2018. *ITUC Frontlines Briefing Climate Justice: COP24 Special Edition.* Brussels: ITUC. www.ituc-csi. org/Frontlines-Briefing-Climate-Justice-Special-Edition-COP24?lang=en (last accessed April 2019).

——. 2017. *Just Transition – Where Are We Now and What's Next? A Guide to National Policies and International Climate Governance.* Brussels: ITUC. www. ituc-csi.org/just-transition-where-are-we-now? (last accessed April 2019).

——. 2016. *Call for Dialogue: Climate Action Requires Just Transition.* Brussels: ITUC. www.ituc-csi.org/call-for-dialogue-climate-action-17837 (last accessed April 2019).

JTC (Just Transition Centre) & The B Team. 2018. *Just Transition: A Business Guide.* www.ituc-csi.org/IMG/pdf/just_transition_-_a_business_guide.pdf (last accessed April 2019).

JTRC (Just Transition Research Collaborative). 2018. *Mapping Just Transition(s) to a Low-Carbon World.* Geneva: Rosa-Luxemburg-Stiftung, University of London Institute in Paris & United Nations Research Institute for Social Development. www.unrisd.org/jtrc-report2018 (last accessed April 2019).

Krieg-Planque, A. 2010. 'La formule "développement durable": Un opérateur de neutralisation de la conflictualité', *Langage et société*, vol. 134, no. 4, pp. 5–29.

Levy, DL & D Egan. 2003. 'A neo-Gramscian approach to corporate political strategy: Conflict and accommodation in the climate change negotiations', *Journal of Management Studies*, vol. 40, no. 4, pp. 803–829.

Milne, MJ, K Kearins & S Walton. 2006. 'Creating adventures in Wonderland: The journey metaphor and environmental sustainability', *Organization*, vol. 13, no. 6, pp. 801–839.

Moussu, N. 2017. 'Entre corporatisme et universalisme: les associations de firmes transnationales face au changement climatique'. PhD dissertation, University of Lausanne.

Newell, P & D Mulvaney. 2013. 'The political economy of the "just transition"', *The Geographical Journal*, vol. 179, no. 2, pp. 132–140.

OECD (Organisation for Economic Co-Operation and Development). 2011. *OECD Guidelines for Multinational Enterprises.* OECD Publishing. www.oecd. org/daf/inv/mne/48004323.pdf (last accessed April 2019).

Paterson, M. 2010. 'Legitimation and accumulation in climate change governance', *New Political Economy*, vol. 15, no. 3, pp. 345–368.

Robins, N, V Brunsting & D Wood. 2018. *Climate Change and the Just Transition. A Guide for Investor Action.* Grantham Research Institute & Initiative on

Responsible Investment, in partnership with the Principles for Responsible Investment and the International Trade Union Confederation. www.lse.ac.uk/GranthamInstitute/wp-content/uploads/2018/12/Climate-change-and-the-just-transition_Guide-for-investor-action.pdf (last accessed April 2019).

Schmidheiny, S (with the Business Council for Sustainable Development). 1992. *Changing Course. A Global Business Perspective on Development and the Environment*. Cambridge, MA & London: The MIT Press.

Snell, D. 2018. "'Just transition'? Conceptual challenges meet stark reality in a "transitioning" coal region in Australia', *Globalizations*, vol. 15, no. 4, pp. 550–564.

TCFD (Task Force on Climate-related Financial Disclosures). 2017. *Recommendations of the Task Force on Climate-related Financial Disclosures*. www.fsb-tcfd.org/publications/final-recommendations-report/ (last accessed April 2019).

The B Team. 2018. *Pledge for a Just Transition to Decent Jobs*. 1 August. www.bteam.org/plan-b/just-transition-pledge/ (last accessed April 2019).

UN (United Nations). 2015. *Transforming Our World: The 2030 Agenda for Sustainable Development*. https://sustainabledevelopment.un.org/post2015/transformingourworld (last accessed April 2019).

UNOHCHR (United Nations Office of the High Commissioner for Human Rights). 2011. *Guiding Principles on Business and Human Rights. Implementing the United Nations 'Protect, Respect and Remedy' Framework*. New York & Geneva: OHCHR. www.ohchr.org/Documents/Publications/GuidingPrinciplesBusinessHR_EN.pdf (last accessed April 2019).

UNFCCC (United Nations Framework Convention on Climate Change). 2015. Adoption of the Paris Agreement. Twenty-first session of the Conference of the Parties. UN Doc. No. FCCC/CP/2015/L.9/Rev.1. 12 December.

USCIB (United States Council for International Business) & BizMEF (Major Economies Business Forum on Energy Security and Climate Change). 2018. *Business Engagement in Implementing National Climate Pledges & The Paris Agreement*. www.uscib.org/uscib-content/uploads/2018/12/2018_11_30_BizMEF_climate_report.pdf (last accessed April 2019).

WBCSD (World Business Council for Sustainable Development). 2018. *Further, Faster: Business Calls For Increased Ambition, Clear Rules and Strong Enablers at COP24*. Geneva: WBCSD. www.wbcsd.org/Programs/Climate-and-Energy/Climate/Climate-Action-and-Policy/News/Further-faster-business-calls-for-increased-ambition-clear-rules-and-strong-enablers-at-COP24 (last accessed April 2019).

——. 2010. *Vision 2050. The New Agenda for Business*. Geneva: WBCSD. www.wbcsd.org/Overview/About-us/Vision2050/Resources/Vision-2050-The-new-agenda-for-business (last accessed April 2019).

——. 1997. *Global Scenarios 2000–2050. Exploring Sustainable Development. Summary Brochure*. Geneva: WBCSD.

We Mean Business. 2018. *Business Calls for Increased Climate Ambition*. www.wemeanbusinesscoalition.org/policy/cop24/ (last accessed April 2019).

——. 2015a. 'The business brief. Shaping a catalytic Paris Agreement', *We Mean Business Blog*, 21 November. www.wemeanbusinesscoalition.org/blog/the-business-brief-shaping-a-catalytic-paris-agreement/ (last accessed April 2019).

——. 2015b. 'This is the turning point', *We Mean Business Blog*, 28 September. www.wemeanbusinesscoalition.org/blog/this-is-the-turning-point/ (last accessed April 2019).

Wei, D. 2018. *Climate and the Just Transition: The Business Case for Action*. San Francisco, CA: Business for Social Responsibility (BSR) & We Mean Business. www.bsr.org/reports/BSR_Climate_Nexus_Just_Transition.pdf (last accessed April 2019).

Wright, C & D Nyberg. 2015. *Climate Change, Capitalism, and Corporations. Processes of Creative Self-Destruction*. Cambridge: Cambridge University Press.

3
Australian business: Embracing, reconceptualising, or ignoring a just transition in Australia

Caleb Goods

INTRODUCTION: CLIMATE CHANGE, BUSINESS AND JUST TRANSITION

Occurring simultaneously on opposite sides of the planet, the 2018 summer heatwave in the northern hemisphere and the devastating drought on the east coast of Australia are glaring reminders of the reality of climate change. In its latest report, the Intergovernmental Panel on Climate Change (IPCC, 2018) explicitly states that unless we undertake swift and decisive action, the heatwaves and droughts of 2018 are set to become the new normal. Climate-related events and efforts to avert catastrophic climate change have profound impacts on how people work, particularly but not exclusively in highly polluting sectors such as coal (Thompson & Richards, 2015). The 2015 heatwave in India, for instance, resulted in taxi unions in Kolkata urging drivers to avoid working between 11 am and 4 pm (ABC, 2015). Recent research has shown that in Cambodia, shifting rainfall patterns, driven by climate change, has led to crop failure and rural worker indebtedness; thus, as a consequence of climate change, workers have been pushed into bonded labour arrangements with brick factory owners (Robertson, 2018).

A steadily growing number of unions and employers' associations recognise both the reality of the climate crisis and the urgent need to act upon it. In some instances, employer groups and unions have worked together and jointly advocated policies to reduce greenhouse gas emissions (Goods, 2017; Markey & McIvor, 2019). Alongside this, a growing body of research in the fields of labour studies, industrial relations and critical management also focuses on the role of unions and businesses. For the most part, the research argues that, for the low-carbon transition to be successful, workers and management must together re-imagine the future of work (see for example, Lipsig-Mumme

& McBride, 2015; Paillé et al., 2014; Wittneben et al., 2012; Wright et al., 2013; Goods, 2017). Among the concepts that are being referred to in discussions on the low-carbon transition involving unions, businesses, policy makers and academics, is just transition, a concept that broadly holds that the shift towards a low-carbon economy can only succeed if it is guided by the principle of equity. In other words, it is the belief that workers and their communities should not disproportionally bear the burden of climate action. Placing justice at the centre of how unions, governments and businesses respond to climate change raises a series of challenges. This is especially true in the business community and even more so in an Australian context, which is marked by widespread climate change agnosticism – an agnosticism that is actively fuelled by a handful of powerful corporate interests.

While a growing body of academic literature focuses on union and environmental justice approaches to just transition (e.g. Stevis & Felli, 2015; Evans & Phelan, 2016), relatively little research has been devoted to how industry and corporate interests engage with the concept, and with questions of climate equity or justice more broadly, and this despite signs of a nascent corporate uptake, especially at the international level (e.g. JTC & The B Team, 2018; Moussu, in this volume). In this chapter, I contribute to filling this gap by exploring the extent to which and if so, how Australian business groups are grappling with the just transition concept and, more generally, issues of work, justice and climate change. My analysis draws on a combination of secondary sources and first-person accounts collected through 22 interviews (conducted between September 2016 and October 2017) with representatives from the main Australian membership-based business associations involved in labour market issues (Sheldon et al., 2019:18). The interviewees were linked to both industry-based – representing particular sectors of the economy such as the resource or retail sectors – and cross-sector business associations.

The chapter begins by setting out the Australian context. Understanding Australia's particular sociopolitical and economic make-up is critical to understanding the country's 'peculiar' relationship with climate change. The Australian context also shapes how industry and business leaders understand and conceptualise climate change. I then go on to investigate the extent to which corporate interests, without always referring to the just transition concept, interpret its underlying principles. From here, the chapter looks at how the organisations frame the just transition concept and the potential implications of this. Having presented the different business conceptualisations of just transition, I put forward three basic heuristics – hardline resistance, corporate reconceptualisation and limited tripartism – for making sense of business engagement with just transition.

JUST TRANSITION ENTERS THE MAINSTREAM

Although the origins of the just transition concept can be traced back to the late 1970s, it has, in recent years, gained growing recognition, largely thanks to the efforts of organised labour. At the international level, this is best exemplified by the efforts of the International Trade Union Confederation (ITUC) and its successful inclusion of the just transition concept in the international climate debate (Burrow, 2015; Rosemberg, in this volume). In December 2015, these efforts resulted in the inclusion of just transition in the preamble of the Paris Agreement on climate change (UNFCCC, 2015). The inclusion of just transition in the Paris Agreement echoes a broader proliferation of the term at international, regional and national levels. Notable examples include the International Labour Organization's publication of its *Guidelines for a Just Transition* (ILO, 2015) or the European Parliament's proposal – following lobbying efforts by the European Trade Union Confederation (ETUC) – to create a Just Transition Fund to help ensure justice for workers and communities impacted by efforts to mitigate greenhouse gas emissions (European Parliament, 2017). Leading environmental organisations such as Greenpeace (Ritter, 2018:27–34) or the World Wide Fund for Nature (WWF, 2016) have also taken up the just transition concept. At the national level as well, we are seeing efforts to include just transition into climate change strategies (e.g. Australian Senate Environment and Communications References Committee, 2017; Government of Canada, 2018).

Over the last decade, the just transition concept has entered the mainstream. At the international level, a growing number of business or business-related groups are recognising this and starting to include the just transition concept in their own climate-related work (see, for example, ICC, 2017). In line with earlier discourses on 'green growth' and the 'green economy', they argue that a just transition is not only beneficial to workers and the environment, but that it also makes 'good business sense'. A handful of academics (Robins et al., 2018) and various international platforms and initiatives – like the Just Transition Centre (JTC) or The B Team (Moussu, in this volume) – promote this 'win-win-win' approach to the just transition. In a joint publication, both the JTC and B Team, for example, argue that 'From a commercial perspective, implementing a just transition allows companies to plan for, manage and optimize the operational and reputational effects of cutting emissions and increasing resource productivity' (JTC & The B Team, 2018:2).

The 'win-win-win' approach, by insisting on the co-benefits of action for businesses, workers and the planet, and positioning businesses as essential drivers of positive change, marks a clear departure from earlier union and environmental justice conceptualisations of just transition (Stevis et al., in this volume). It also raises a number of more general questions about the underlying agendas, worldviews and understandings of what the word 'justice' actually stands for.

Are we witnessing a similar trend at the Australian level? Are Australian business groupings also embracing/appropriating the just transition concept and pushing a 'win-win-win' agenda of their own? As was previously highlighted, Australia represents an extreme case, given that a number of prominent Australian business groupings and politicians continue to actively promote a 'pro-fossil fuel' policy and economic agenda. In order to understand the Australian business community's position with respect to climate change and just transition, it is necessary to make sense of its particular relation to fossil fuels – and most notably, coal.

AUSTRALIA AND CLIMATE CHANGE:
A FRACTIOUS RELATIONSHIP

For Australia to ratify the Kyoto Protocol is to commit to a policy of decarbonisation of our industrial and export base. It is to inflict upon its people unemployment, profound economic dislocation and decline. For a nation to inflict upon itself such economic and political damage is almost unprecedented.

Australian Prime Minister (1996–2007)
John Howard (Howard, 2006:67)

… one of the greatest challenges we face is climate change and the threat it poses to the environment. In December last year, after many years delay, Australia ratified the Kyoto Protocol. We are determined to be part of the solution to climate change not just a part of the problem.

Australian Prime Minister (2007–10 & 2013)
Kevin Rudd (Rudd, 2008)

Both quotes provide a glimpse of the political, economic and ideological tensions that have surrounded the climate issue in Australia over the last thirty years. The Australian Government's first real attempt at reducing its greenhouse gas emissions, heavily caveated by economic prerequisites, can be traced back to the late 1980s (Lowe, 1994). Since then, Australia's successive energy and climate policies have been characterised by their unpredictability and haphazardness. This has had very negative impacts on workers and local communities (Sheldon et al., 2018). Two interlocking factors explain this fractious relationship with the climate issue. The first relates to the particular historical trajectory of Australia's political economy. And the second relates to the extraordinary political power and influence wielded by the vested interests that disproportionally benefitted from this trajectory.

Australia is a resource-rich, fossil fuel export-oriented economy. In the financial year 2016–17, the Australian resources sector generated export revenues of approximately AUS\$ 198 billion, a figure that amounts to more than the total revenues of all other industries combined (MCA, 2018). Since the mid-twentieth century, exports have essentially targeted Asian markets that are heavily reliant on Australian minerals to fuel their own economic development (Ellem & Tonts, 2018). In particular, China's insatiable hunger for Australian iron ore and coal has pumped billions of dollars into global mining companies such as BHP and Rio Tinto, and into the treasuries of successive Australian governments (Ellem & Tonts, 2018).

This financial windfall places resource companies and the industry associations that represent them in a powerful position. Ellem (2016), for example, shows how the pursuit of a neoliberal agenda directly responded to the demands and interests of Australia's mining and resource sector. The same applies to climate policy. Pearse (2007) convincingly describes the powerful role that these highly polluting industries have played in shaping the Australian Government's policy response to climate change. In particular, he looks at how Australian governments have regularly included lobbyists from the resource and mining industries in their delegations at international climate negotiations.

For a group of influential political and economic leaders, coal exports, and more specifically coal for domestic baseload energy production, has come to symbolise Australia's past and future prosperity. So much so that in 2017, Australia's then Treasurer, and now Prime Minister, Scott Morrison carried a lump of coal into the Australian Parliament and, coal in hand, declared:

This is coal. Do not be afraid. Do not be scared. It will not hurt you … It is coal that has ensured for over 100 years that Australia has enjoyed an energy-competitive advantage that has delivered prosperity to Australian businesses and has ensured that Australian industry has been able to remain competitive in a global market. (Morrison, 2017:536)

This political economic context has converted and cemented many Australian political, business and union leaders to what Levy & Spicer (2013) call a 'fossil fuels forever' mindset of what is in Australia's best interests. 'Fossil fuels forever' feeds on the idea that the science of climate change is not settled and that controls on fossil fuel exploration and emissions would cause unnecessary economic disruption, threatening the future prosperity of Australian businesses and quality of life of its citizens. It is actively mobilised by employer associations from the resource sector and has enabled them to exert great influence over Australian policy and public discourse (Murray, 2006; Pusey, 1991), and especially in the areas of climate and energy policy (Curran, 2011; Evans & Phelan, 2016). A potent example is the plan of Adani Mining – a subsidiary of India's Adani Group – for a coalmine in the Galilee Basin in Central Queensland. According to estimates, the mine could

… produce up to 330 million tonnes per year of coal for export, releasing an estimated 705 million tonnes of carbon dioxide per year (more than Australia's total annual domestic greenhouse gas emissions) that would contribute to a catastrophic global warming trajectory of up to 6 degrees Celsius. (Ritter, 2018:6)

Despite the overwhelming arguments against it, Adani's plans have received backing from the current Australian Government and various business and union groups.

The entrenchment of the 'fossil fuels forever' mindset has also shaped the views and strategies of those who believe in the climate science and in the need to take action – especially within the business community. As one NGO representative I interviewed told me, the presence or absence of climate change in the political debate 'depends upon what the political view of the day is and people's willingness to engage in things' (interview 14). A representative from an employer association not connected to the resources sector, expressed his frustration at this 'fossil fuel forever' mindset and the need to keep discussions of climate action vague in the Australian context:

There is a whole segment of the coalition [government] that is just not interested in listening … For someone who likes to believe in the power of evidence, reason and argument, it is pretty bloody depressing to see what has happened in the last decade. We have had more reports and modelling exercises and evidence examining what to do about climate change than we have had on just about anything else and we have not gotten anywhere in terms of persuading people.

People that were sympathetic remain sympathetic and people who are unsympathetic are immovable. (Interview 8).

The same person then went on to describe how Australia's 'fossil fuels forever' mindset complicated the task of coming up with policy proposals that would both provide a meaningful business response to the climate challenge and at the same time be acceptable to its membership:

> I think the biggest challenge is that 'tribe' has taken over much more than philosophy or ideology, in how climate change is engaged with by different people. Within our membership I come across individuals at companies from time to time who are fervently – vehemently – full spectrum sceptical [of climate change]. But it has almost never got anything to do with their business, it is a personal thing, and they are embedded within a wider tribal identity that incorporates climate scepticism as one of its tenets and tests of purity. So that crops up more as an issue in trying to get policy and political solutions because you might think that all business groups, right of centre groups, will listen to business and business groups and they kind of don't at least on issues where they think it goes to their core identity. (Interview 8)

The power of the fossil fuel and extractive industries, the political class's fear and/or commitment to these industries, and the constraints the 'fossil fuels forever' mindset imposes on those who wish to act on climate change, are central to understanding why Australia has, for the last thirty years, been unable to take decisive action on climate. While Australia is a signatory to the 2015 Paris Agreement, it has so far failed to take any decisive steps to meet its international carbon reduction commitments. In fact, Australia's greenhouse gas emissions have actually risen since 2015 (Slezak, 2018). It is against this backdrop that Australian business groups have been engaging with the climate debate and just transition concept.

AUSTRALIAN BUSINESSES AND CLIMATE CHANGE

Given the nature of Australia's political economic context, climate change is a conflict-ridden topic within the Australian business community. Some business leaders, for instance, continue to question the climate science, arguing that it remains unsettled. Others, without necessarily questioning the science, express concern about the potentially negative effects of climate policies on their businesses' bottom lines. In other words, regardless of one's personal views on the issue, dealing with climate change essentially boils down to making sure that a given industry and/ or business is not adversely affected by climate-related policies. For representatives from the resource sector, who are, as I have shown, powerful and influential players in the Australian policy debate, the priority is to make sure that climate policy does not get in the way of business as usual.

It should be noted that some, within the industry and business community, agree with the science and recognise the threats posed by a rapidly changing climate. They also acknowledge that a degree of government intervention is required in order to get businesses to change their practices. As one interviewee explained, policy consistency and a clear sense of direction from government is paramount in order for businesses to adapt and plan accordingly (interview 1). In various cases, these calls were coupled with calls for financial incentives for businesses – through tax cuts and funding for innovation. As another interviewee explained, 'I would say that 95 per cent of our members don't care about climate change. They do however care about the bottom line. And so because they care about the bottom line, how would climate change responses help their bottom line?' (interview 4).

As the interviews show, business attitudes towards the climate issue are a reflection of broader attitudes towards government action. There was a clear division between business leaders who saw government action on climate change as a danger and those who felt that a degree of government intervention was necessary in order to secure their business interests. These differences notwithstanding, there is widespread acknowledgment, and albeit for diametrically opposite reasons, that climate change is set to remain on the political and business agenda, and that Australian businesses can therefore not afford to ignore it. In a 2018 survey of over 1,200 Australian company directors, climate change ranked as the number one issue that businesses felt needed government attention (Khadem, 2018). These different attitudes towards climate change and climate policy have both led business representatives to engage with the just transition issue, and simultaneously shaped their attitudes towards it.

BUSINESS ENGAGEMENT WITH JUST TRANSITION IN AUSTRALIA

In their article 'Global Labour Unions and Just Transition to a Green Economy', Dimitris Stevis and Romain Felli (2015) associate the just transition concept with three main justice principles: redistribution, recognition and participation. The principle of redistribution, they argue, seeks to guarantee that the burden of transition is fairly distributed across society. Recognition involves making sure that workers and their communities – especially those that are directly affected by mitigation and adaptation efforts – are explicitly accounted for in government and corporate efforts to address the climate challenge. Finally, the principle of participation means getting all sections of society to participate in the planning and undertaking of the transition towards a low-carbon economy – including unions and representatives from industry. Drawing on Stevis and Felli's useful categorisation, I explore, in the following pages, the extent to which Australian businesses have taken up the just transition concept in their climate-related work. To what extent and how are they mobilising the three justice principles identified by Stevis and Felli? As the analysis shows, Australian business groups' attitudes towards just transition are not homogenous. They can, in fact, be classified in three broad categories: hardline resistance, corporate reconceptualisation and limited tripartism. What follows is an exploration of each of these three business conceptualisations of just transition.

Hardline resistance

As has already been outlined, a sizeable amount of Australian businesses – as well as noteworthy sections of society and the political establishment – actively resist the idea that Australia, and the world, should engage in any sort of transition towards a low-carbon economy. Consequentially, the very *idea* of a just transition becomes superfluous. The Minerals Council of Australia (MCA), the leading industry group representing Australia and other major mining and energy interests (including BHP, Rio Tinto, BP, Peabody Energy and Glencore), best encapsulates this idea. In a recent submission to an Australian Senate inquiry examining the just transition of industry, workers and communities in Australia's coal-based electricity generation sector, particularly in the Latrobe Valley (see Snell, in this volume), the MCA outlined its rejection of the premise for a just transition in the following way:

> Instead of thinking black and brown coal's day has passed and focussing on the structural adjustment pressures resulting regionally from closure of generators and industrial plants, the Inquiry needs to bear in mind coal's potential to support new industries and jobs in those same communities well into the future. More widely, across the entire economy, access to affordable and reliable energy is an essential underpinning of growth and our standard of living. (MCA, 2016:4)

From the perspective of business leaders and groups such as the MCA, justice involves protecting existing highly polluting industries such as coal and the jobs associated with them – and conceivably greening them through technological fixes like 'clean coal'. It should be noted that MCA's position is, however, subject to growing scrutiny and criticism as shareholders of some of its most powerful members – such as BHP – have started to openly criticise its hardline resistance to climate action (Thomson, 2017).

There exists, within the business community, a very different, but equally significant, form of hardline resistance to just transition and related principles, this time focusing on the negative social effects of the low-carbon transition. While acknowledging the need to confront climate change, some corporate interests argue that the disruption and social dislocation caused by mitigation efforts is an unfortunate but necessary price to pay to secure a low-carbon future. As one business leader put it, 'unfortunately in some instances, and look at Latrobe Valley and the other analogy would be automobile industry manufacturing, *sometimes it's just hard luck* ... So it is always terrible when you see that happen but it does happen with structural change' (interview 18, emphasis added).

Hardline resistance to just transition dominates the Australian business community. For the most part, it either reflects a continued commitment to the 'fossil fuel forever' mentality or a Schumpeterian (Schumpeter, 1950) belief that disruption – and its social implications – is a normal and necessary part of overall prosperity. Any form of state or corporate intervention to 'soften the blow' is therefore unwarranted. Either way, this ultimately means rejecting the very idea of a just transition.

Corporate reconceptualisation

In their article 'An Inconvenient Truth: How Organizations Translate Climate Change into Business as Usual', Christopher Wright and Daniel Nyberg have convincingly shown how organisations are adept at internalising and subsequently converting the 'grand challenge' of climate change into the 'mundane and comfortable concerns of "business as usual"' (2017:1633). This process of appropriation and reconceptualisation also applies to certain business attitudes to just transition in Australia. In the instances where business groups have discussed and used the just transition concept, it has often led to the reinterpretation of the ideas and principles associated with the term in a way that serves their short-term interests – rather than the common good.

In its submission to the Australian Senate inquiry on coal-fired power generation and just transition, the Australian Mines and Metals Association (AMMA), for instance, stated that 'The resource industry supports genuine and constructive policy debate on Australia's energy future to ensure an appropriate and Just Transition for industry, employees and communities' (2016:1). The AMMA then goes on to write that 'Justice is not solely about money or alternative employment/opportunity; a genuinely Just Transition must also provide certainty and predictability' for business (2016:15). The focus on certainty and predictability within the AMMA's definition of just transition is in line with the 'mundane and comfortable concerns of "business as usual"' mentioned above. This commitment to a 'business as usual' approach becomes even clearer when it associates just transition with energy security, national economic prosperity and job creation and retention:

> ... if Australian miners (including coal miners) were unable to secure reliable and affordable sources of energy (electricity), this would cause significant harm and jeopardize some proportion of over 200,000 direct mining jobs ... Not only does Australia gain economically from having a strong export commodity base ... but Australia also has a humanitarian role to play in exporting a comprehensive suite of energy commodities (including coal, renewables, gas, nuclear – strictly for power generation only) to energy importing nations. (AMMA, 2016:2–3)

As one interviewee from the resource sector put it:

> ... initially employer associations generally say fight – it is a horrible word – with unions, but in this case we are in the same boat. We want people to have jobs, respectable jobs and reasonably well paid jobs. Now obviously as a result of these significant changes, the outcome, the social outcome, I mean a higher renewable energy economy will have lesser jobs, I mean that is just the outcome. (Interview 3)

The corporate reconceptualisation approach essentially involves securing and even increasing profit margins under the guise of job protection and economic development at home and abroad. By reaffirming the importance of fossil fuels alongside sustainable energy options, it actually includes very little transition at all but rather upholds – albeit behind a green veil – the 'fossil fuels forever' approach. In short, the corporate reconceptualisation approach is grounded on a very narrow and flawed understanding of the transition that, while accepting the reality of climate change, ultimately ends up reviving the 'jobs vs environment' dichotomy (Räthzel & Uzzell, 2011) – a dichotomy that early proponents of just transition specifically sought to abolish (Goods, 2013).

Limited tripartism

The final approach to just transition identified within the Australian business community involves recognising the scale of the climate challenge and the need for a degree of collaboration, albeit on business terms. It acknowledges the need for governments and businesses to support workers in the transition to a low-carbon economy. As one business leader explained, 'one of the issues that we spent a half day on was just transition … we do need to make some kind of arrangements here' (interview 8). Just transition is presented as making good business sense, especially given the growing preoccupation with climate change among certain sections of the workforce. As another business representative put it, 'people don't want to work for a business that has no interest in these areas. So I think it is really hard to draw really good talent from the gen y's [people born between 1980 and 1994] – who we want in our business' (interview 11).

While some business leaders acknowledge the need to address workers' concerns, they are reluctant to offer them a seat at the table. Only one – out of the 22 interviewees – openly supported joint union-employer action on climate change through social dialogue and improved workplace relations. The overwhelming majority opposed worker participation, arguing that it was the management's role to decide, and workers' role to implement. As one interviewee put it, 'look, at the end of the day the responsibility sits with the business' (interview 1). He went on to argue that ideas of social dialogue or worker consultation were European imports, ill-suited to the Australian context:

> Things like, particularly in Western Europe, those very strong employee consultative committees have been a feature for some time. It certainly has been a significant issue of focus for them for managing their industrial relations for an awful long period of time, and unions are embedded in that process. It has a very different feel to how we operate. There has been a number of attempts by different organisations to try and transplant that approach to Australia and it does not work. (Interview 1)

For a number of business representatives I interviewed, ensuring a just transition was less about engaging workers at the company level than getting the national and state governments to create the right conditions – and especially regulatory environment – for the transition to take place. As one interviewee noted, above all, what most business leaders are asking for is that the 'government creates a stable [and] effective carbon regulatory regime' (interview 18).

If we refer back to Stevis and Felli (2015), the limited tripartism approach is therefore fundamentally about ensuring a smooth, business-friendly transition towards a low-carbon economy. While this involves a degree of recognition of workers' concerns and some, albeit limited, concessions to secure worker buy-in, it does not entail offering a voice to workers or empowering them. Unsurprisingly, the overriding priority, as in both other approaches, remains the bottom line.

MAKING SENSE OF THE THREE BUSINESS APPROACHES

A combination of factors explains why Australian business representatives – including 'progressive' business leaders – have either completely dismissed the just transition idea or acknowledged its relevance while simultaneously emptying it of its transformative or empowering potential. First, business representatives' overall dismissal of worker participation within the context of a just transition echoes a more general, overarching trend. In Australia, over the last thirty years, we have seen a movement away from a limited form of industrial democracy, active worker voice and participation, towards individualism and the growing assertion of managerial prerogatives (Veen, 2018; Cooper & Ellem, 2008). The very idea that workers and their unions should be consulted and involved in the shift towards a green economy is inconceivable for most Australian business leaders.

Second, as outlined at the beginning of this chapter (and acknowledged by most business leaders interviewed), climate change is a deeply toxic issue in Australian politics and society. This contributes to root climate scepticism within the Australian business community and creates few incentives for serious action. More specifically, the volatility and lack of cross-party commitment to climate action makes it hard for many business leaders to even conceive of the need for a just transition – and there is little prospect for this to change in the short to medium term. Not surprisingly, one of the fundamental reasons why successive Australian governments have failed to act on climate change is because of fierce opposition and lobbying from business, particularly those involved in the fossil fuel industry.

Third, when business leaders do acknowledge the need for a just transition, it is premised on the prioritisation of their business interests – the pursuit of profit above all else – rather than the interests of society as a whole. One progressive business leader interviewed in the research project summed up the way in which this tension between climate risk and financial risk mediates the business world's engagement with climate change:

> They [business leaders] also have a fiduciary duty. Everyone is looking at fiduciary duty … If you do not manage climate change risk you could be exposed, that is the headline. But the other way of looking at fiduciary duty is that you cannot just make materially significant business decisions without an expectation that it is going to deliver a return to your shareholders over a period of time … So they have that balance, their fundamental purpose is to deliver outcomes for shareholders. (Interview 18)

In short, business representatives' approach to the climate issue and stance in relation to just transition is exclusively driven by the management of risk and the maximising of profits. This can mean rejecting the just transition concept altogether (hardline resistance), re-appropriating and reconceptualising it by stripping it of its progressive elements (corporate reconceptualisation), or accepting a limited and controlled level of worker involvement, as long as the overriding interests of the company are secured.

The more regressive approaches to just transition – hardline resistance and corporate reconceptualisation – are the most prevalent, or at least visible, positions within Australian business. There is, however, reason to believe that the balance of power may be – albeit very slowly – shifting towards limited tripartism (Khadem, 2018). The fact that previously hostile business groups (Hart, 2008) in the fossil fuel industry are now calling on the Australian Government to put a price on carbon (Morgan, 2018) is an indicator of this. So is the Victorian State Government's decision to support just transition initiatives in the coal-fired power generation sector. As to whether these and other measures will lead to a greater consideration of climate justice by business remains to be seen. What is certain, however, is that unions, especially given their relative power in a number of key high-emitting sectors, will have a critical role to play in pressuring corporations and businesses to meaningfully engage with just transition in a way that does not just imply 'business as usual'.

CONCLUSION

An examination of Australian business leaders' engagement with climate change and more specifically just transition reveals that when Australian businesses express a willingness to take up the just transition concept, they do so on their own terms and in a way that strips it of its transformative potential. The toxicity of climate politics in Australia has a lot to do with this, a toxicity that was produced and is upheld by the very same Australian business groups and leaders. Resistance, reconceptualisation and limited acceptance of just transition from Australian business leaders is therefore unsurprising. Nevertheless, there are three aspects from the research findings that I believe are of relevance beyond the Australian context and are worthy of particular consideration.

First, arguably the more destructive dimension of hardline business resistance to just transition is its rejection in the name of neoliberal ideology, since it involves a full-blown rejection of state intervention – through industry policy, community revitalisation, or worker compensation and training – and worker participation. The pursuit of just transition must therefore overcome an orthodoxy that has in many parts of the world been deceitfully presented as 'common sense'. In Australia, challenging the 'common-sense' rhetoric is proving particularly difficult as thirty years of hardline neoliberalism have stripped state institutions of their capacity to act – a point made clear by the shortcomings of green stimulus programmes rolled out in Australia after the global financial crisis (Goods, 2011). There is a need to both counter neoliberal assertions against just transition's core tenets and build capacity to implement a justice-based green political economy shift.

Second, the research underscores that there is a new frontier in corporate greenwashing strategies, namely 'just transition vs the environment' that aims to divide and weaken the counterhegemonic forces that are seeking to embed a just transition in a radical, transformative agenda. Just transition is presented as a choice between workers' economic needs on the one hand, and the environment on the other. Again, this is not new, but its implications are strong – pitting workers, their unions and communities against each other and the environment, is counter to the very spirit of just transition and thus signals a 'just transition gone wrong'.

Third, as noted at the beginning of this chapter, unions and workers have been adept at promoting the just transition concept, if not embedding its actual practice, which has led to a growing engagement with business and points to an inherent tension. The growing appeal of the just transition concept provides it with the power to unite, even with some sections of the business community, as highlighted in this research, and drive transformative change. Yet, as this research also shows, just transition's growing appeal leaves it vulnerable to the demands and influence of powerful political economic forces whose underlying agendas are fundamentally at odds with just transition or restrict it to rare 'win-win-win' outcomes for business, workers and the environment. This situation is exacerbated by the declining role and power of unions and workers across many countries. While businesses and business leaders must be part of the transition, a just transition should not be about prioritising their financial interests over those of workers and the environment. For a low-carbon transition to be just, it should first and foremost ensure justice for workers and their communities.

REFERENCES

ABC (Australian Broadcasting Corporation). 2015. 'India heatwave: More than 500 dead as temperatures near 50 degrees Celsius', *ABC*, 26 May. www.abc.net.au/news/2015-05-25/india-heatwave-kills-hundreds/6496314 (last accessed April 2019).

AMMA (Australian Mines and Metals Association). 2016. *Submission to the Senate Environment and Communications References Committee Inquiry into the Retirement of Coal-fired Power Stations*. Canberra: Australian Parliament.

Australian Senate Environment and Communications References Committee. 2017. *Senate Inquiry: Retirement of Coal Fired Power Stations*. Canberra: Commonwealth of Australia.

Burrow, S. 2015. 'A call to action: Climate & secure jobs', Speech to the Climate Change Conference, Oslo, 13 March 2015. International Trade Union Confederation. www.ituc-csi.org/sharan-burrow-s-speech-to-the?lang=en (last accessed April 2019).

Cooper, R & B Ellem. 2008. 'The neoliberal state, trade unions and collective bargaining in Australia', *British Journal of Industrial Relations*, vol. 46, no. 3, pp. 532–554.

Curran, G. 2011. 'Modernising climate policy in Australia: Climate narratives and the undoing of a Prime Minister', *Environment and Planning C: Government and Policy*, vol. 29, no. 6., pp. 1004–1017.

Ellem, B. 2016. 'Geographies of the labour process: Automation and the spatiality of mining', *Work, Employment and Society*, vol. 30, no. 6, pp. 932–948.

—— & M Tonts. 2018. 'The global commodities boom and the reshaping of regional economies: The Australian experience', *Australian Geographer*, vol. 49, no. 3, pp. 383–395. doi: 10.1080/00049182.2017.1388761.

European Parliament. 2017. 'MEPs back plans to cut carbon emission allowances and fund low-carbon innovation', 15 February. www.europarl.europa.eu/news/en/press-room/20170210IPR61806/meps-back-plans-to-cut-carbon-emission-allowances-and-fund-low-carbon-innovation (last accessed April 2019).

Evans, G & L Phelan. 2016. 'Transition to a post-carbon society: Linking environmental justice and just transition discourses', *Energy Policy*, vol. 99, pp. 329–339.

Goods, C. 2017. 'Climate change and employment relations', *Journal of Industrial Relations*, vol. 59, no. 5, pp. 670–679.

——. 2013. 'A just transition to a green economy: Evaluating the response of Australian unions', *Australian Bulletin of Labour*, vol. 39, no. 2, pp. 13–33.

——. 2011. 'Labour unions, the environment and green jobs', *Journal of Australian Political Economy*, vol. 67, June, pp. 47–67.

Government of Canada. 2018. *Just Transition Task Force.* 16 February. www.canada.ca/en/environment-climate-change/news/2018/02/just_transition_taskforce.html (last accessed April 2019).

Hart, C. 2008. 'ETS dead: Woodside – The long slow crash of '08', *The Australian*, 14 October.

Howard, J. 2006. *House of Representatives: Hansard Forty First Parliament*, 2 November. Australian Government.

ILO (International Labour Organization). 2015. *Guidelines for a Just Transition Towards Environmentally Sustainable Economies and Societies for All.* Geneva: ILO.

ICC (International Chamber of Commerce). 2017. *ICC Key Messages COP23 2017.* https://cdn.iccwbo.org/content/uploads/sites/3/2017/11/cop23-key-messages-01.pdf (last accessed April 2019).

IPCC (Intergovernmental Panel on Climate Change). 2018. 'Summary For policymakers', in V. Masson-Delmotte et al. (eds), *Global Warming of 1.5 °C. An IPCC Special Report on the impacts of global warming of 1.5°C above pre-industrial levels and related global greenhouse gas emission pathways, in the context of strengthening the global response to the threat of climate change, sustainable development, and efforts to eradicate poverty.* Geneva: World Meteorological Organization.

JTC (Just Transition Centre) & The B Team. 2018. *Just Transition: A Business Guide.* www.ituc-csi.org/IMG/pdf/just_transition_-_a_business_guide.pdf (last accessed April 2019).

Khadem, N. 2018. 'Why Australian company directors have started caring about climate change', *Australian Broadcasting Corporation*, 24 October. www.abc.net.au/news/2018-10-25/why-company-directors-have-started-caring-about-climate-change/10423658?section=business (last accessed April 2019).

Levy, D & A Spicer. 2013. 'Contested imaginaries and the cultural political economy of climate change', *Organization*, vol. 20, no. 5, pp. 659–678. doi: 10.1177/1350508413489816.

Lipsig-Mumme, C & S McBride. 2015. *Working in a Warming World.* Kingston: McGill-Queens University Press.

Lowe, I. 1994. 'The greenhouse effect and the politics of long-term issues', in S Bell & B Head (eds), *State, Economy and Public Policy in Australia*, pp. 315–333. Melbourne: Oxford University Press.

Markey, R & J McIvor. 2019. 'Environmental bargaining in Australia', *Journal of Industrial Relations*, vol. 61, no. 1, pp. 79–104.

MCA (Minerals Council of Australia). 2018. *Submission to Resources 2030 Taskforce.* Canberra: Australian Government.

——. 2016. *Submission to the Inquiry into the Retirement of Coal Fired Power Stations*. Canberra: Australian Government.

Morgan, E. 2018. 'Woodside boss Peter Coleman calls for Australia to introduce a carbon price', *Australian Broadcasting Corporation*, 14 November. www.abc.net.au/news/2018-11-14/woodside-ceo-peter-coleman-argues-for-carbon-price/10494026 (last accessed April 2019).

Morrison, S. 2017. *House of Representatives Official Hansard*. Canberra: Commonwealth of Australia.

Murray, G. 2006. *Capitalist Networks and Social Power in Australia and New Zealand*. Farnham: Ashgate.

Paillé, P, Y Chen, O Boiral & J Jin. 2014. 'The impact of human resource management on environmental performance: An employee-level study', *Journal of Business Ethics*, vol. 121, no. 3, pp. 451–466.

Pearse, G. 2007. *High & Dry*. Camberwell: Penguin Books.

Pusey, M. 1991. *Economic Rationalism in Canberra: A Nation-building State Changes Its Mind*. Cambridge: Cambridge University Press.

Räthzel, N & D Uzzell. 2011. 'Trade unions and climate change: The jobs versus environment dilemma', *Global Environmental Change*, vol. 21, no. 4, pp. 1215–1223.

Ritter, D. 2018. *The Coal Truth: The Fight to Stop Adani, Defeat the Big Polluters and Reclaim our Democracy*. Crawley: UWA Publishing.

Robertson, H. 2018. '"Blood bricks": How climate change is trapping Cambodians in modern slavery', *Australian Broadcasting Corporation*, 16 October. www.abc.net.au/news/2018-10-16/how-climate-change-is-trapping-cambodians-into-modern-slavery/10377982 (last accessed April 2019).

Robins, N, V Brunsting & D Wood. 2018. *Investing in a Just Transition: Why Investors Need to Integrate a Social Dimension Into Their Climate Strategies and How They Could Take Action*. The Centre for Climate Change Economics and Policy. www.lse.ac.uk/GranthamInstitute/wp-content/uploads/2018/06/Robins-et-al_Investing-in-a-Just-Transition.pdf (last accessed April 2019).

Rudd, K. 2008. 'Speech by Prime Minister Kevin Rudd to the United Nations General Assembly', 26 September. www.un.org/ga/63/generaldebate/pdf/australia_en.pdf (last accessed April 2019).

Schumpeter, JA. 1950. *Capitalism, Socialism and Democracy*. New York: Harper and Row.

Sheldon, P, E Della Torre & R Nacamulli. 2019. 'When territory matters: Employer associations and changing collective goods strategies', *Human Resource Management Journal*, vol. 29, no. 1, pp. 17–35.

Sheldon, P, R Junankar & A De Rosa Pontello. 2018. *The Ruhr or Appalachia? Deciding the Future of Australia's Coal Power Workers and Communities*. Sydney: Industrial Relations Research Centre. www.ituc-csi.org/IMG/pdf/ruhrorappalachia_report_final.pdf (last accessed April 2019).

Slezak, M. 2018. 'Australia's emissions rise again in 2017, putting Paris targets in doubt', *The Guardian*, 29 March. www.theguardian.com/environment/2018/mar/29/australias-emissions-rise-again-in-2017-putting-paris-targets-in-doubt (last accessed April 2019).

Stevis, D & R Felli. 2015. 'Global labour unions and just transition to a green economy', *International Environmental Agreements: Politics, Law and Economics*, vol. 15, no. 1, pp. 29–43.

Thompson, G & D Richards. 2015. 'The end of coal?' *Australian Broadcasting Corporation*, 16 June. www.abc.net.au/4corners/stories/2015/06/15/4253096. htm.

Thomson, J. 2017. 'BHP to review Minerals Council membership after green push', *The Financial Review*, 19 September. www.afr.com/business/mining/bhp-to-review-minerals-council-membership-after-green-push-20170919-gykfjl (last accessed April 2019).

UNFCCC (United Nations Framework Convention on Climate Change). 2015. *Adoption of the Paris Agreement. Twenty-first session of the Conference of the Parties*. UN Doc. No. FCCC/CP/2015/L.9/Rev.1. 12 December.

Veen, A. 2018. 'The bulwarking of individualised workplace relations in the mining industry: Retaining managerial control within a contested regulatory space', *Journal of Industrial Relations*, vol. 60, no. 1, pp. 3–52.

Wittneben, BBF, C Okereke, SB Banerjee & DL Levy. 2012. 'Climate change and the emergence of new organizational landscapes', *Organization Studies*, vol. 33, no. 11, pp. 1431–1450.

WWF (World Wide Fund for Nature). 2016. 'Stay the course: How do we remain on track between Paris and Marrakech? WWF "asks" for UNFCCC intersessional, Bonn, May 2016'. https://d2ouvy59p0dg6k.cloudfront.net/downloads/wwf_asks_for_bonn_intersessional__may_2016_.pdf (last accessed April 2019).

Wright, C & D Nyberg. 2017. 'An inconvenient truth: How organizations translate climate change into business as usual', *Academy of Management Journal*, vol. 60, no. 5, pp. 1633–1661.

Wright, C, D Nyberg, C De Cock & G Whiteman. 2013. 'Future imaginings: Organizing in response to climate change', *Organization*, vol. 20, no. 5, pp. 647–658.

4
Tales from the frontlines: Building a people-led just transition in Jackson, Mississippi

Kali Akuno

INTRODUCTION

There's no Hispanic air, no African American air, or white air, there's just air. And if you breathe air, and most people I know do breathe air, then that makes you part of the environment and if you are concerned about the quality of that air, I would consider you an environmentalist. And if you drink water, and most people I know drink water, and you are concerned about what's in the water, then I would consider you an environmentalist. And you eat food, and again most people I know eat food, and you are concerned about what's in the food, then I would consider you an environmentalist. If you answer two of the three, then I would say you are an environmentalist, you just might not know it.

These wise words are those of Robert Bullard, a leading academic and campaigner against environmental racism (1990). They perfectly capture the spirit and ambition of Cooperation Jackson: organising and empowering ordinary people, particularly in Black and Latinx communities, by connecting their day-to-day struggles and concerns – breathing clean air, drinking non-polluted water, eating food that is both healthy and plentiful – to the broader systemic – social, economic, environmental – crises that we collectively face.

Launched on 1 May 2014, Cooperation Jackson presents itself as a vehicle to build a vibrant, ecologically regenerative and democratic solidarity economy in Jackson, Mississippi while simultaneously anchoring these efforts within a broader ecosocialist project. Cooperation Jackson's basic theory of change is centred on the idea that organising and empowering the structurally under- and unemployed sections of the working class, particularly from Black and Latinx communities, will be a catalyst for the democratisation of our economy and society in general. The core vehicle through which to attain these objectives is the construction of an interconnected and self-sustaining network of green worker-owned cooperatives and supporting institutions. With a little over a hundred dues-paying members, its mission is not only to assist Jackson residents but to also empower them by providing them with an entirely new, supportive economy in which to operate. As Peter Moskowitz wrote, 'The idea is essentially this: since Jackson's current economy isn't working for its residents, and its current political system isn't doing much to help, why not create a new economic and political system right alongside the old one?' (Moskowitz, 2017).

Among the words and concepts that Cooperation Jackson uses to both describe and organise its efforts is that of just transition. Its understanding of just transition, and its approach to realising it, builds on a long history of oppression and struggle in a city, state and country that is marred by social, environmental and racial inequality and injustice. It draws inspiration from a variety of ideas and projects, reinterpreting and adapting them to the particular economic, political and social realities of Jackson. Securing a just transition is therefore about empowering and organising frontline communities, developing autonomous, self-reliant spaces of power, while simultaneously pursuing a political goal, and engaging in municipal, state, federal and even international political spaces (especially through our membership in alliances such as the Climate Justice Alliance). As Beppe Caccia notes:

> What is most inspiring about [Cooperation Jackson] is its ability to combine in a long-term process a radical vision inspired by grassroots initiatives with a comprehensive plan of socio-economic transition; and then and finally producing the political change necessary at a local level to implement the general plan. (Transformative Cities Initiative, 2018)

It is about simultaneously and inextricably transitioning the self, the community, the city and the world. In this sense, it differs from more common understandings of just transition that centre on the energy sector and workers, and that do not carry a broader, emancipatory and revolutionary political project.

JACKSON, MISSISSIPPI: A CITY IN CRISIS

As Jackson's late Mayor Chokwe Lumumba explains, 'Jackson, like many urban centres, is struggling to overcome decades of economic divestment, deindustrialization, suburban flight, a declining tax base, chronic under and unemployment, poorly performing schools, and an anti-quated and decaying infrastructure' (Lumumba, 2014). With a median household income of just under US$40,000, Mississippi is the poorest state in the US. With a population of approximately 170,000, Jackson, the state's capital, is worse off still. The median household income in Jackson amounts to just over US$33,000 per year. For the city's people of colour, who are the majority of the city's population, the situation is even worse. While approximately 14 per cent of Jackson's White residents are below the poverty line, the figure grows to 34 per cent in the case of Black residents.

High poverty rates are coupled to economic – and particularly indus-trial – decline and underfunded/waning municipal and state services. As the capital of one of the most conservative states in the country, Jackson has been subjected to countless state policies that have almost exclu-sively served the interests of its richest residents. These include some of the largest tax breaks in Mississippi history, which have contributed to further reducing the overall tax base and thus depriving municipal-ities and public agencies of the minimum resources needed to provide Jackson's residents – especially in the poor, predominantly Black, neigh-bourhoods – with the most basic of public services such as schools and clean running water. One Republican state legislator even floated plans for a Detroit-style emergency management bill that would allow the gov-ernor's office to take control of city services, starting with its municipal airport (Nave, 2016).

To this should be added Jackson's vulnerability to environmental and climate-related risks and hazards – especially health-related issues – that disproportionately affect the city's poorer and majority Black neighbour-hoods. In parts of South and West Jackson, for instance, the proximity of residential areas to highways and polluting industries is linked to chronic levels of asthma and other breathing and heart-related problems (see NRDC, 2017). The city is also home to numerous toxin-contami-nated sites. While a number of these have been 'officially' recognised as 'brownfield sites', a great many more have not as a result of inadequate government monitoring (which, in turn, is the result of the commitment of insufficient resources). These sites are usually residential in nature, or situated near residential zones or schools, and therefore pose devel-opmental threats for our children and carcinogenic risks more broadly.[1] When we combine these local sources of pollution with the broader

---

1   For information on identified brownfield sites in Jackson, MS, see http://www.homefacts.com/environmentalhazards/superfunds/Mississippi/Hinds-County/Jackson.html (last accessed June 2019).

effects of climate change, such as rising temperatures, it is no surprise that Jackson was named the Allergy Capital of the United States in 2016 by the Asthma and Allergy Foundation of America (Ginty, 2017).

The city's water management issues are becoming legendary. Not only is the water unfit for human consumption, but it also acts as a transmitter of communicable diseases. In 2016, tests showed that in over 22 per cent of Jackson homes, lead levels in the water exceeded the federal 'action' level. This, it was shown, was due to the inadequate monitoring of pipe corrosion by the local authorities. More worryingly, government officials knowingly waited over six months before issuing a public warning (Galbraith & Teague, 2016). Jackson's water issues are long-standing and directly related to intentional neglect and institutional and systemic racism. The neglect of the system began when the city transitioned from being majority White to majority Black, without any corresponding transfer of the city's wealth and means of production.

Another issue of concern, which also overwhelmingly affects the poorer and majority Black communities, relates to city waste management and illegal dumping, which is not only a source of pollution and contamination, but also a major producer of carbon and methane gases, two major contributors to climate change. Jackson is also one of the largest contributors to climate change in the state of Mississippi as a direct result of how it receives and consumes its energy (The White House Office of the Press Secretary, 2014), and the presence of large polluting industries in and around the city that rely on trucking, railroad and airfreight transportation.

LOOKING BACK: COOPERATION JACKSON'S APPROACH TO JUST TRANSITION

From its inception in May 2014, the just transition framework has formed a central feature of Cooperation Jackson's strategy and efforts to address the aforementioned and deep-rooted problems. The concept is at the centre of the Sustainable Communities Initiative, a comprehensive roadmap developed by Cooperation Jackson whose goal is to protect the environment, curb carbon emissions, stimulate employment, and democratically transfer wealth and equity at the municipal level.

Cooperation Jackson has developed a holistic and radical approach to just transition. This involves imagining, developing and implementing very concrete, grassroots-led and focused initiatives, while simultaneously pursuing a broader, transformative ecosocialist agenda. Its members are bound by a shared worldview and overarching goal that involves radically shifting wealth and power, and overhauling how things are produced, distributed, consumed and recycled back into the natural resource systems that we depend on. Having said this, it also believes in the need to ground these ideas in everyday struggles and practice. As a result, Cooperation Jackson constantly adapts its programme and strategy so as to factor in new challenges and seize upon new opportunities (Akuno, 2017:3).

At the frontline community level, Cooperation Jackson focuses on advancing the social and solidarity economy. This serves a broader political purpose as an instrument of 'popular education' that provides local communities with a clear and tangible sense of the broader issues at stake and what they can achieve through collective mobilisation. In addition to responding to the urgent needs of frontline communities, locally grounded initiatives contribute to culture change. They shape human behaviours – particularly the behaviours that guide our collective choices about who decides what we produce and consume, why we produce and consume, and why what we produce and consume is distributed in the inequitable manner that it is. Transformative change requires cultural transformation. As brandon king[2] explains,

> … it takes us taking steps away from the TV screen and actually seeing each other, being with each other, being in community with each other … And this is something we have to relearn … To be completely 100, all this stuff we're doing? We're learning while doing … It's being the example and showing the alternative – I think when people see it, and they see how much fun we're having, that draws folks to it. (Harvey, 2018)

Cooperation Jackson's just transition efforts draw on a long, rich and at times painful, history of resistance and struggle. In particular, its 'roots lay deep within the struggle for democratic rights, economic justice, self-determination, particularly for Afrikan people in the Deep South' (Cooperation Jackson, n.d.). Jackson forms a historic battleground for civil rights struggles in the United States. It hosted a number of emblematic civil rights protests in the 1950s and 1960s and it was also the site of tragic events, such as the assassination in 1963 of Medgar Evers, a prominent civil rights activist.

Its founding members are, for the most part, associated with locally based movements, community groups and political formations. These include the Republic of New Afrika movement (Cunningen, 1999; for a history of black nationalism, see van Deburg, 1997; Haywood, 1978;

---

2   He voluntarily spells his name in lower case.

Jeffries, 2006), the Malcolm X Grassroots Movement (MXGM), as well as the Jackson People's Assembly and Mississippi Disaster Relief Coalition, both of which were created in the aftermath of Hurricane Katrina (August 2005). An important and inspirational local figure is Chokwe Lumumba, a long-time activist and leader of the Republic of New Afrika movement. Following a career in human rights law in Detroit, Lumumba returned to Jackson in the 1980s to engage in grass-roots organising and municipal politics. With the active support of MXGM (which he helped found), the Mississippi Disaster Relief Coalition, and a range of other community groups, Lumumba was elected to the Jackson Ward Two council seat in the 2009 Municipal Elections. He then went on to win the Mayoralty in 2013 before prematurely dying in early 2014.

In the lead-up to the 2009 and 2013 elections, Lumumba consistently centred his campaigns on environmental justice concerns. He relentlessly argued that it was possible to simultaneously address Jackson's particular exposure to environmental and climate-related risks while dealing with its social and economic difficulties through the creation of sustainable jobs and cooperative enterprises. His environmental justice focus was shaped by his and his supporters' first-hand experience of Hurricane Katrina (2005) and its aftermath. Hurricane Katrina's disastrous effects on New Orleans are widely acknowledged, but the story of its impact on Mississippi is far less known. While over 50 per cent of the Gulf Coast region, from Texas to the Florida Panhandle, was directly affected by the hurricane, its effects were hardest felt in southern Mississippi where the eye of the storm landed. Gulfport, Biloxi and countless other towns, cities and hamlets along the Mississippi Gulf Coast suffered as much, if not greater, damage than New Orleans. Many of the smaller towns have still not fully recovered. Several of Cooperation Jackson's core members, including myself, Sacajawea Hall and brandon king, were actively involved in the grass-roots relief and recovery efforts in New Orleans and the Mississippi Gulf Coast from September 2005 through April 2008. This difficult experience profoundly shaped our own collective consciousness of the severity of the climate crisis, who was the most exposed to it, and why it was so urgent to act. Many of us, including myself, subsequently became avowed and committed ecosocialists.

FROM THE JACKSON-KUSH PLAN TO COOPERATION JACKSON AND THE JUST TRANSITION PLAN

Developed over the 2004–12 period by a work-study group of the New Afrikan People's Organization and the Malcolm X Grassroots Movement, the Jackson-Kush Plan formed the programmatic backbone of Chokwe Lumumba's electoral campaigns in 2008–09 and 2012–13. In addition to supporting and organising an independent electoral force, it also involves building people power through people's assemblies and the social and solidarity economy. As we will see, a number of the ideas contained in the plan informed Cooperation Jackson's Sustainable Communities Initiative and just transition strategy. Through its three programmatic components – building people's assemblies, building an independent black political party, and building a broad-based solidarity economy – the plan's ambition was to provide a compelling and comprehensive roadmap to achieve autonomous power in Jackson, Black self-determination and the democratic transformation of the economy. It was about developing 'a mass base with the political clarity, organizational capacity, and material self-sufficiency to advance the objective of building an autonomous power' (Akuno, 2012:4).

While stressing the importance of building autonomous power outside of the state structures (through people's assemblies), the J-K Plan also stresses the importance of engaging in electoral politics on a limited scale 'with the express intent of building radical voting blocs and electing candidates drawn from the ranks of the Assemblies themselves'. As the plan stipulates, 'as we learned through our own struggles – and through our analysis of the experiences of many other revolutionary or liberation movements – we ignore the power of the state at our own peril' (Akuno, 2012:7). Political engagement is seen as a means to

> ... create political openings that provide a broader platform for future struggles to be waged to restore the 'commons', to create more public utilities (i.e. universal health care and comprehensive public transportation), and for the democratic transformation of the economy ... While we believe [in] building autonomous power outside of the state (i.e. government) structures ... we also believe that engaging electoral politics on a limited scale with the express intent of building radical voting blocs and electing candidates drawn from the ranks of the Assemblies themselves is important. (Ibid.)

With the passing of Mayor Chokwe Lumumba in early 2014, the primary responsibility for developing and carrying out the Jackson-Kush Plan fell on the shoulders of Cooperation Jackson. The decision to create Cooperation Jackson was also influenced by changes in the broader municipal political context. While initially presenting himself as a worthy successor of Lumumba and a supporter of the J-K Plan, the new mayor, Tony Yarber (elected in April 2014), finally chose to ignore its core elements on the grounds that the city could not afford them. Given the municipal administration's reluctance to act, the pro-Lumumba movements and the community groups that were behind Cooperation Jackson decided to press on with the J-K Plan by focusing on its 'solidarity economy' elements. In other words, adverse political circumstances did not represent a barrier for action. In fact, Cooperation Jackson takes pride in not depending on electoral cycles to move things forward (this comes back to its broader understanding of institutional politics as a means to an end rather than an end in itself).

For Cooperation Jackson, the priority therefore became the advancement of its plans for a local economy grounded on the principles of social solidarity, mutual aid, reciprocity and generosity – priorities that, as we have seen, formed a pillar of the Jackson-Kush Plan. From a practical standpoint, it involved developing a network of cooperative institutions that serve the community and promote the aforementioned principles. These range from worker cooperatives to informal affinity-based neighbourhood bartering networks.

A wide range of experiments and initiatives in other regions and countries inspired Cooperation Jackson – and the movements that are behind it – to formulate what the 'solidarity economy' could look like. Influences include earlier efforts by the New Afrikan people to develop cooperatives, as well as Latin American social movements' opposition to neoliberal and neocolonial austerity measures in the 1980s and 1990s. An important source of inspiration is the Mondragon Federation of Cooperative Enterprises based in the Basque region of Spain. Through its 261 businesses, 101 of which are cooperatives, and with its 75,000 employees, Mondragon is regularly presented as an example of the transformative potential of worker cooperatives.

As previously mentioned, far from being an end in themselves, cooperatives are seen as a means to an end. They form part of a wider transitional strategy and praxis to advance the abolition of capitalism and combat the poverty and oppressive social relations that it fosters. The development of cooperatives serves to heighten the contradictions within the capitalist system, a political approach that sets Cooperation Jackson apart from most other cooperative projects. There are plenty of examples of well-run economic alternatives to dominant economies, from worker-run laundromats in Cleveland to the countless food coops throughout the United States. However, many shy away from explicitly political goals. Cooperation Jackson, by contrast, wants to use worker-owned and managed companies to create an economic alternative for Black Jackson residents outside of capitalism, and moreover in an era of conservative political rule.

It was shortly after its creation that Cooperation Jackson and its associated movements began to more systemically refer to the 'just transition' concept. This coincides in particular with growing uses of the concept by the US environmental justice movement. The Climate Justice Alliance, for instance, of which Cooperation Jackson is a member, launched a national campaign on the topic in 2013: the 'Our Power Campaign: Communities United for a Just Transition' (Smith & Patterson, 2019). The just transition concept increasingly served as a common ground for political and strategic conversations within the environmental justice community and between environmental justice groups and other groups – especially trade unions (see Stevis et al., in this volume). These intensified in the lead-up to the Paris Climate Conference in December 2015 which led to the inclusion of just transition in the preamble of the Paris Agreement on climate change (see Stevis et al., in this volume).

The just transition concept also formed the basis for deep conversations within Cooperation Jackson. It was following a year of intense discussions, that Cooperation Jackson came up with its own 'Just Transition Plan' which, by acting as a kind of roadmap for a just, low-carbon transition, was intended to guide Cooperation Jackson's efforts at the municipal level and to strengthen alliances with other local constituencies. Interestingly, it is worth mentioning that the plan received support from local trade unions (the United Automobile Workers and the Mississippi Alliance of State Employees/Communications Workers of America) and progressive 'social gospel' churches. Drawing on the feedback from partners and community groups, Cooperation Jackson issued a new comprehensive 'working draft' version of the Just Transition Plan in September 2016. Its launching was strategically timed to shape the debate and narrative around the 2017 municipal elections, and to get Mayoral and City Council candidates to commit to adopting the plan should they take office – and in particular to get the Yarber administration to adopt it. The Just Transition Plan's overall objective is:

> To improve the quality of life in our City and for the sake of our children, grandchildren, and great-grandchildren we can and must end the overlapping environmental, climatic and human rights crises confronting us. Cooperation Jackson believes that we can solve these crises by organizing our communities to execute a comprehensive program that will protect our environment, curb our carbon emissions, stimulate employment, and democratically transfer wealth and equity. (Cooperation Jackson, 2015).

It is premised on putting an end to Jackson's 'systemic dependence on the hydro-carbon industry and the capitalist driven need for endless growth on a planet with limited resources, while creating a new, democratic economy that is centred around sustainable methods of production and distribution that are more localized and cooperatively owned and controlled' (Cooperation Jackson, 2015). In addition to its broader goals, the plan also contains very concrete objectives. These include developing a network of green cooperatives, building an eco-village, and, at the municipal level, getting the administration to adopt and implement just transition policies.

Having imagined and developed an alternative model, the greatest challenge for Cooperation Jackson became of implementing it on the ground. In the following section, I provide a critical account of Cooperation Jackson's efforts to translate the Just Transition Plan into reality.

JUST TRANSITION IN PRACTICE

To what extent has Cooperation Jackson succeeded in implementing its vision of a just transition? As we have just noted, a core component of the plan involves creating a green cooperative ecosystem that reinforces and builds upon itself. In its short existence, Cooperation Jackson has successfully launched several functioning and emerging cooperatives and acquired a community centre and twenty parcels of land in West Jackson. Practical initiatives include Freedom Farms, an urban farming cooperative; Nubia's Place and Catering Cooperative, a café, and the Green Team, a landscaping, organic waste-gathering and composting cooperative. A core purpose of these cooperatives is to demonstrate in practice what a just transition can look like, as well as help counter the reactionary forces of gentrification and displacement.

Based in West Jackson, Freedom Farms currently grows produce on two acres of land owned by the community-controlled Fannie Lou Hamer Community Land Trust. This cooperative urban farm specialises in organic vegetables and is currently expanding its production to fruits and aquaculture. Most of Freedom Farm's produce is sold to Nubia's Place Café and Catering Cooperative, a 'healthy-oriented' worker-owned café and catering service, whose purpose is to fight chronic obesity and the diabetes-related afflictions that threaten the lives of many local residents. The Green Team, a yard and lawn care, landscaping and composting worker-owned cooperative, processes the organic waste generated by Nubia's Place. It also gathers organic materials from grocery stores and restaurants and turns this organic waste into compost that is sold to gardeners, farmers and hardware and home supply stores. Through these three cooperatives, the idea was to create a reinforcing value chain wherein Freedom Farms produces food that is sold and consumed at Nubia's Place Café, the waste from which is utilised by the Green Team to create organic compost that nourishes the crops produced by Freedom Farms. In addition to these three cooperatives, and to meet the essential objective of owning and controlling the means of production, Cooperation Jackson created a Community Production Cooperative, a small-scale manufacturing cooperative specialising in the arts and science of digital fabrication. With this technology in hand, we aim to fashion new, innovative ways of moving towards carbon-neutral and fully recyclable production.

In line with our 'circular economy' approach, our aim is to install our first digitally fabricated housing and communal structures within a planned eco-village (the Ewing Street Eco-Village Pilot Project). The eco-village project's ambition is to create a sustainable live-work community in West Jackson. Through permaculture-based landscaping design, the eco-village will be built on land protected by the Community Land Trust. Through its digitally fabricated housing and communal facilities and the renovation of two existing structures (on the block currently in the Community Land Trust holding), the eco-village will, by late 2019, provide affordable cooperative housing and jobs through a number of integrated and interdependent cooperative enterprises that will be situated within the community. These will include jobs in urban farms, composting operations, childcare, solar-thermal installation and maintenance, security, arts and culture, and a grocery store. The eco component of the community includes the creation of an integrated solar-thermal, recycling, and composting network in the community. This will provide affordable and sustainable energy and green jobs that will help fight ecological degradation and climate change. The exercise of collective land and home ownership, and the provision of permanent affordability and tax controls, will enable us to fight the encroachment of gentrification and displacement threatening the predominantly Black working-class community of West Jackson.

Other future plans include expanding our Community Land Trust with the purchase and first-stage development of the Ida B. Wells Plaza, and turning the grocery unit of the plaza into a multipurpose space that will house a) a hydroponic and aquaponic growing operation, b) a café and catering operation, c) a venue to host local artisans, and d) a venue to host weekly and/or monthly farmers' markets and swap meets. Together, these existing and projected cooperative units contribute to advancing several of our regenerative economic objectives: regenerating our soils to serve as effective carbon sinks, building the zero-waste initiative by expanding our composting practices, and developing the skill and capacity to move towards food security as the first step towards attaining food sovereignty.

MUNICIPAL POLITICS

While grass-roots organising, base building, and worker-owned cooperative development are essential components of Cooperation Jackson's just transition work, we also believe in the need to engage with institutional politics and to actively pressure the municipal government to implement citywide policies that curb ecological destruction and climate change and incentivise the creation of sustainable jobs and cooperative enterprises. Cooperation Jackson is, in this sense, committed to helping the city realise the vision of the short-lived Lumumba administration: turning Jackson into the most 'sustainable city' in the South. More specifically, this means getting the city to commit to becoming a zero-emissions and zero-waste city by 2025. To attain both objectives, Cooperation Jackson is pressing the city government to implement the following policies:

1. Retrofitting and weatherising all of the buildings owned and operated by the city, so that they conserve heat in the winter and naturally cool the facility in the summer. We also want the city to incentivise this type of retrofitting in the private and nonprofit sectors of the economy with grants, low-interest loans, tax-credits, etc.
2. Place solar panels on all municipal buildings and facilities that have the capacity to host the equipment. We also want to encourage the city to install solar-thermal converters in all of the facilities it possesses that have the capacity to regulate their energy use via this technology. Furthermore, we want the city to incentivise private and nonsector solar-thermal energy conversion and production and enable residents and businesses to supply excess energy to the main power grid to aid the energy company's efforts to eliminate its dependence on fossil fuels.
3. Gradually replace the municipality's entire operating fleet, including all of police vehicles, with electric vehicles. We want to encourage the city to incentivise the purchasing of electric cars and to create publicly owned and operated electric fuelling stations throughout the city to accommodate this transition.
4. Develop an expanded and sustainable public transportation system. We want to push the City of Jackson to gradually acquire

a fully electric public transportation fleet and to expand its public transportation vehicles, routes and hours to accommodate more efficient and accessible transportation throughout the city and metro-region.

5. Implement a comprehensive recycling programme, that includes mass public education, and a system of inducements and rewards for residents, businesses and civil institutions in the city to recycle all that can be recycled to reduce the burden on the city's landfill and to create more private- and public-sector jobs in waste management and recycling;

6. Create a comprehensive composting programme that gathers all of the organic refuse produced by households, businesses and civil institutions and include the requisite public education necessary to encourage individuals, families, businesses and institutions to participate and to adhere to all of the necessary sanitary standards.

7. Create a comprehensive cooking oil-gathering programme that calls for all restaurants and food service businesses and institutions producing mass amounts of used cooking oils for their food production such as schools, colleges, universities, and hospitals to recycle these materials so they can be reused for other energy and production needs and help eliminate the need for their extended production and disposal at public expense.

8. Create a Local Food and Production Charter, to encourage and incentivise local food production and distribution, to create more jobs and reduce carbon emissions by eliminating the need for extended transportation systems and refrigeration. The incentive programme should focus exclusively on supporting producers who reside in Jackson and are drawn from historically discriminated and capital-deprived communities.

Through Cooperation Jackson's involvement in larger activist networks and alliances, policy efforts at the local level are combined with campaigning activities at the US national level, especially in relation to the emerging discussions on the Green New Deal.

NATIONAL POLITICS: COOPERATION JACKSON AND THE GREEN NEW DEAL

In the US, the current debate on the Green New Deal provides a historic opportunity for environmental justice groups to advance their understandings of what a just transition should look like. The proposed congressional resolution, submitted by Democratic Congresswoman

Alexandria Ocasio-Cortez and Senator Ed Markey, explicitly refers to the concept and calls, among other things, for the '[achievement of] net-zero gas emissions through a fair and just transition for all communities and workers' (House Resolution 109, 2019). Through its involvement in the It Takes Roots Alliance – a coalition of alliances including the Climate Justice Alliance, the Grassroots Global Justice Alliance, the Indigenous Environmental Network, and the Right to the City Alliance (RTC) – Cooperation Jackson plans on playing a critical role in helping to shape the Green New Deal going forward. As with our previous efforts at the municipal level, we firmly believe in the need to work within that political space, not to be part of it, but rather to create the conditions for the ecosocialist future we need. This means making sure that this critical piece of legislation is firmly rooted in a just transition framework that, in turn, reflects our own, our allies', and more broadly frontline communities' concerns.

To the extent that it is rooted in a radical just transition framework, the Green New Deal has the potential of significantly contributing to solving the climate crisis – especially when we consider the United States' disproportionate and historic responsibility in this crisis. It can potentially provide a blueprint for a different economy grounded on sustainable, collectively owned and democratically controlled systems of production, distribution and consumption. But a well-crafted plan and large amounts of federal funding will not be enough. The Green New Deal's success will also hinge on its ability to radically improve the everyday lives of the frontline communities most exposed to the climate threat. In other words, the Green New Deal will have to be sufficiently nuanced and adjustable to the multiplicity of situated racial and class injustices that currently cut across US society. It must therefore not be unidimensional in its orientation – i.e. only concerned with federal or state-level emissions reductions – but it must also factor in and overcome the racial, class, gender, and regional-based injustices that are deeply embedded in US society.

For this to happen, we must make sure that federal dollars earmarked for the Green New Deal *actually* reach the right people and are used in a way that is relevant and adjusted to their needs. This, among other things, means bypassing state legislatures and creating direct channels of communication and exchange between, on the one hand, frontline groups and, on the other, the federal government and the agencies responsible for implementing the deal. This is especially important in a historically Republican state such as Mississippi where, as was previously highlighted, successive state governments have systematically ignored poor, majority Black and environmentally vulnerable communities in cities such as Jackson. Under present political conditions, if Green New Deal funds were allocated to the state government and the state government alone, frontline communities in West Jackson would only end up receiving a fraction of the money they require.

To be effective, the Green New Deal must also place questions of reparations, decolonisation and indigenous sovereignty at its core. On the issue of reparations, this means unequivocally changing our collective relation to the lands that provide us with the sustenance that we need to live. A starting point should be the restoration of indigenous sovereignty and engaging in a real, concrete process of decolonisation. Reparations should also involve a massive transfer of resources to people of African descent.

While people's livelihoods and the climate are key, the Green New Deal should also address the critical issue of natural habitat loss. The sixth mass extinction in earth's history, currently under way, is marked by an unprecedented decline in natural habitats and the corresponding annihilation of wildlife. In addition to playing a key role in the preservation of biodiversity, natural habitats also act as natural carbon sinks and as barriers and regulatory mechanisms in the case of natural hazards such as floods. Protecting and restoring natural habitats will require new productive approaches and techniques, especially in sectors such as agriculture. The role of cities and urban spaces will be especially important. We will need to maximise urban density to allow for the preservation and restoration of natural habitats (replenishment of the soil, restoration of ecological balance, recuperation of species). Urban spaces can also act as 'living farms' to address many of our caloric needs. The Green New Deal is going to have to address this challenge head on and leave ample room for experimentation, but an experimentation that intentionally breaks the power of the corporate monopolies and creates new incentives for production that is not profit-driven or -bound.

Finally, the Green New Deal, while focusing on the US context, should also be firmly grounded in an internationalist perspective. This is critical given the global nature of the climate crisis and the global reach of the US economy. The GND should consequently include a ban on mining, and petrochemical, agricultural and fishing operations by US multinationals in developing regions (Africa, Asia, the Caribbean, Latin America and Oceania). It should also promote the development of open-source technologies to directly transfer technology and information to peoples throughout the world, thereby allowing communities to produce the new carbon-reducing or carbon-neutral technologies they need. And finally, it should include policies that eliminate the impositions by the World Trade Organization that negate national and local sovereignty, and which have prevented the introduction of major climate mitigation initiatives in the US and Canada.

Through this short – and incomplete – overview, we seek to highlight how the Green New Deal offers a historic opportunity to, once and for all, set us on a path towards a more just, low-carbon world. I have also insisted on the questions that it raises and that will need to be addressed if we are to ensure that the Green New Deal actually delivers the real-life effects that are so desperately needed to forestall climate catastrophe.

CHALLENGES AND ROADBLOCKS

While Cooperation Jackson's just transition efforts have produced tangible results, it has faced and continues to face a series of challenges and roadblocks, especially in relation to its broader political goal. The first relates to the cooperative economy dimension of our work. The worker owners of Nubia's Place Café and Catering Cooperative, which was one of the three initial worker-run cooperatives launched by Cooperation Jackson, decided to become a wholly independent entity for political reasons. In particular, they no longer wished to align themselves with Cooperation Jackson's collective political outlook and worldview, particularly the Jackson Just Transition Plan and our commitment towards ecosocialism from the ground up. What lessons can be learnt from this decision? And in particular, what does this tell us about the inherent challenges associated with efforts to break away from the capitalist logic and capitalist social relations? One of the core underlying reasons for breaking away from Cooperation Jackson was financial in nature: operating in a sustainable and regenerative fashion reduces the profit margins of the business. While they expressed their continued commitment to recycling and reuse, they were less prepared to make the required sacrifices necessary to promote ecosocialism through the development of a non-exploitative and regenerative system at the community level.

The splintering off of Nubia's Place Café points to a second broader and ongoing challenge: changing cultures and shifting individual mindsets and behaviours, and getting people to connect their everyday forms of resistance to a broader systemic outlook and political consciousness. A series of critical steps must be taken before we get to the level of mass direct action on an ongoing and consistent basis. We have to do a much more thorough job of getting people to understand the severity of the crisis *and* our collective ability to do something about it. We have to win over hearts and minds; and we have to defeat this notion that capitalism cannot be overcome. We must do more to convince people en masse that preserving life on this planet will require us to engage in an epic struggle against forces of reaction that are doing everything in their power to undermine our efforts and preserve the status quo – as we have seen over the last decade from the Movement for Justice for Oscar Grant, to Occupy, to the Movement for Black Lives, to Standing Rock.

The difficulties encountered at the frontline community level are reflective of a broader reluctance by our political leaders – most notably at the municipal level – to implement the policy proposals contained in either the Jackson-Kush Plan or the Jackson Just Transition Plan. These proposals, like the Green New Deal at the federal level, may seem radical but are in fact very sensible given the scale of the social and climate challenges we face. As we have previously argued, the success of these and any other plans will hinge on our ability to produce a 'cultural shift' within the local population. Change, whether it be in policy circles or at the frontline community level, will only materialise if it is consciously driven by the people. This is why, since its launch in May 2014, Cooperation Jackson places so much emphasis on developing a bottom-up, participatory approach to just transition. Changing cultures and mindsets is an arduous, grinding, and often unglamorous task that requires patience and commitment. It involves organising – rather than mobilising – people: building a relationship with them based on respect and learning about our shared interests, jointly working to transform our interests into a programme of action, building the capacity to implement our programme, and building an organisation to enable our collective capacity to be exercised to produce the outcomes we desire. This means building relationships and engaging frontline communities through community-focused initiatives that address their day-to-day concerns. Only then can we create the conditions for the cultural shift that we so desperately need.

This type of work can take years, if not decades to accomplish. Given the urgency of the situation, it can be tempting to only focus on the easy victories and low-hanging fruit, and to neglect the more political and transformative mid- to long-term objectives. We can sometimes be fooled into thinking that mobilising people – getting them to sign an online petition, or to take part in a march – will be enough to bring about transformative change. While this is a step in the right direction, it will not be enough to travel the distance to radical change.

WAYS AND MEANS: BY WAY OF CONCLUSION

To achieve a just transition at the local or global level, be it through the Green New Deal, the municipal government, or frontline community efforts, we must organise. Nothing can be achieved unless we organise a strong independent base to advance the transition programme we need. Without independent organisation, the epic challenge of climate change will be held hostage to forces seeking to maintain the capitalist status quo.

Organising requires simultaneous efforts at different levels and in different spaces. We have to organise a mass social movement base within the working class, particularly around the job-focused dimension of just transition. Labour unions, worker cooperatives, workers' centres and other labour-focused groups are key to organising workers and pushing for a programme that addresses workers' immediate and mid-term needs for jobs and stable incomes through the expansion of existing 'green' industries and the development of new ones. One of the initiatives that we as Cooperation Jackson are arguing for is the development of a broad 'union-co-op' alliance that would seek to unite different organised working-class movements in this country around what we call a 'build and fight' programme (Akuno, 2017).

As Cooperation Jackson, we and our allies in the environmental and climate justice community, have consistently argued, we believe in the need to organise frontline, poor and often majority Black and Latinx communities who are both the least responsible for the climate crisis and the most exposed to its consequences. And finally, we also believe in the need to connect our local and national struggles with those of other communities across the globe – especially in the Global South – by adopting an internationalist outlook and organising global networks of solidarity and resistance.

From our perspective, and as we have seen, this involves constructing new worker-owned and self-managed enterprises rooted in sustainable methods of production on the build side, and finding ways for workers to appropriate and to manage the means of production on the fight side, with the goal of transitioning workplaces and industries towards sustainable practices (or in some cases phase them out completely). We believe that this will allow us to build the independence and power we need to dictate the terms of the political struggle in the electoral arena. As was previously stated, we are convinced that in order for a transformative just transition to materialise, this mass social movement base must understand and relate to electoral politics – at the local, state and national levels – as a strategic tool, and not an end in itself.

The devastating landfall of Hurricane Katrina on 29 August 2005 was a tragic wake-up call. Climate change and the threat that it poses to humanity and nature suddenly became a tangible and harsh reality. The creation of Cooperation Jackson and the development of a just transition plan represents our best effort thus far towards halting the looming climate catastrophe and building an ecosocialist future from the ground up. Errors have been made and a lot still needs to be done, within a very short timeframe. So, let's get to it, we have no time to waste.

Ecosocialism or barbarism. The choice is ours.

REFERENCES

Akuno, K. 2017. 'Build and fight: The program and strategy of Cooperation Jackson', in K Akuno & A Nangwaya (eds), *Jackson Rising: The Struggle for Economic Democracy and Black Self-determination in Jackson, Mississippi*, pp. 3–41. Montreal: Daraja Press.

——. 2012. *The Jackson-Kush-Plan: The Struggle for Black Self-Determination and Economic Democracy.* Malcolm X Grassroots Movement & New Afrikan People's Organization. www.scribd.com/doc/218031046/The-Jackson-Kush-Plan (last accessed April 2019).

Bullard, R. 1990. *Dumping in Dixie: Race, Class, and Environmental Quality.* Boulder, CO: Westview Press.

Cooperation Jackson. 2015. *The Jackson Just Transition Plan. A Vision of Cooperation Jackson and the Our Power Campaign to make Jackson (MS) a 'Sustainable City'.* https://cooperationjackson.org/blog/2015/11/10/the-jackson-just-transition-plan (last accessed April 2019).

——. n.d. *The Story of Cooperation Jackson.* https://cooperationjackson.org/story (last accessed 23 April 2019).

Cunningen, D. 1999. 'Bringing the revolution down home: The Republic of New Africa in Mississippi', *Sociological Spectrum*, vol.19, no. 1, pp. 63–92.

Galbraith, K & M Teague. 2016. 'High levels of lead found in Mississippi capital's water likened to Flint crisis', *The Guardian*, 17 March. www.theguardian.com/us-news/2016/mar/17/high-levels-lead-mississippi-water-flint-michigan (last accessed April 2019).

Ginty, MM. 2017. 'Climate change is in the air', *NRDC Personal Action*, 11 July. www.nrdc.org/stories/climate-change-air (last accessed April 2019).

Harvey, S. 2018. 'Leave No Worker Behind. Will the just transition movement survive mainstream adoption?' *Earth Island Journal*, Summer 2018. www.earthisland.org/journal/index.php/magazine/entry/leave_no_worker_behind/ (last accessed April 2019).

Haywood, H. 1978. *Black Bolshevik: Autobiography of an Afro-American Communist.* Chicago, IL: Liberator Press.

House Resolution 109. 2019. *Recognizing the Duty of the Federal Government to Create a Green New Deal.* 116th Congress. www.congress.gov/bill/116th-congress/house-resolution/109 (last accessed April 2019).

Jeffries, JL. 2006. *Black Power in the Belly of the Beast.* Chicago, IL: University of Illinois Press.

Lumumba, C. 2014. *Jackson Rising: Building the City of the Future Today.* https://jacksonrising.wordpress.com/local/jackson-rising-statement/ (last accessed April 2019).

Moskowitz, P. 2017. 'Meet the radical workers' cooperative growing in the heart of the Deep South', *The Nation*, 24 April. www.thenation.com/article/meet-the-radical-workers-cooperative-growing-in-the-heart-of-the-deep-south/ (last accessed April 2019).

Nave, RL. 2016. 'The battle for the Jackson Airport: Can the state take it away?' *Jackson Free Press*, 6 January. www.jacksonfreepress.com/news/2016/jan/06/battle-jackson-airport-can-state-take-it-away/ (last accessed April 2019).

NRDC (Natural Resources Defense Council). 2017. *Climate Change and Health. Air Quality.* www.nrdc.org/climate-change-and-health-air-quality#/map?cities=show (last accessed April 2019).

Smith J & J Patterson. 2019. 'Global climate justice activism: "The New Protagonists" and their projects for a just transition', in R Frey, P Gellert & H Dahms (eds), *Ecologically Unequal Exchange. Environmental Injustice in Comparative and Historical Perspective*, pp. 245–272. London: Palgrave Macmillan.

The White House Office of the Press Secretary. 2014. *FACT SHEET: What Climate Change Means for Mississippi and the Southeast and Caribbean.* https://obamawhitehouse.archives.gov/sites/default/files/docs/state-reports/MISSISSIPPI_NCA_2014.pdf (last accessed April 2019).

Transformative Cities Initiative. 2018. 'People power drives social and ecological transition: Cooperation Jackson', *Atlas of Utopias.* https://transformativecities.org/atlas-of-utopias/atlas-31/ (last accessed April 2019).

Van Deburg, WL. 1997. *Modern Black Nationalism: From Marcus Garvey to Louis Farrakhan.* New York: New York University Press.

5
What transition? Collectively imagining a just and low-carbon future for Río Negro, Argentina

Martín Álvarez Mullally, Fernando Cabrera Christiansen
& Laura Maffei

INTRODUCTION[1]

In its concluding observations on Argentina's fourth periodic report (November 2018), the UN Committee on Economic, Social and Cultural Rights expressed its 'concern about plans for large-scale exploitation of unconventional fossil fuels (shale gas end shale oil) through hydraulic fracturing in the Vaca Muerta region' arguing that it 'would consume a significant percentage of the entire global carbon budget for achieving the 1.5°C target laid down in the Paris Agreement on climate change.' It went on to recommend that Argentina

> ... reconsider the large-scale exploitation of unconventional fossil fuels through hydraulic fracturing in the Vaca Muerta region [and] encourages the State party to promote alternative and renewable energy sources, reduce greenhouse gas emissions and set national targets with time-bound benchmarks (UN, 2018)

The UN Committee is referring to the current Argentinian government's decision to encourage the development of hydraulic fracturing – or fracking – in Northern Patagonia; and this regardless of the growing recognition that averting catastrophic climate change will require a speedy phasing out of conventional and unconventional fossil fuels. While highlighting the potentially devastating effects of the Vaca Muerta plans for the climate, the UN Committee does not mention the disastrous social and economic consequences for workers and local communities. Indeed,

---

1 This chapter was translated from its original Spanish version by Edouard Morena.

the Vaca Muerta megaproject has already affected the Neuquen province and will radically transform the local economy of Río Negro, which, up to recently, was centred on relatively low-carbon productive activities – most notably agriculture.

In response to the government's plans, a group of local activists and concerned citizens set up a *Mesa de Transición Post Petrolera* (Post Petroleum Transition Roundtable – hereafter Mesa) in Río Negro in May 2018. Around thirty people (from Río Negro, Neuquén and Buenos Aires) were actively involved in organising public meetings in Fiske Menuco on 20 July 2018 and 28 September 2018 (Gral. Roca, Río Negro) to critically discuss the Vaca Muerta megaproject. Participants included academics – mostly from the National University of Comahue and the National University of Río Negro – trade union representatives – particularly from the Association of State Workers and the Union of Education Workers of Río Negro (UnTER) that have a long history of involvement in these issues – and concerned citizens and activists involved in the local socioenvironmental movement. The Mesa's ambition is not to organise or coordinate socioenvironmental struggles but rather to share knowledge and draw on existing social forces to come up with an alternative future to the one that the national and provincial authorities are currently offering. Its starting point is that the Río Negro province is currently at a historical juncture and has the opportunity to choose between two very different directions. It can either transition towards an uncertain and unsustainable future centred on unconventional hydrocarbon exploitation, or it can choose an alternative, people-centred just transition pathway that is socially and environmentally sustainable. As the following working hypothesis indicates, the Mesa has clearly opted for the latter:

The promotion of hydrocarbon activities at the provincial level is a response to the economic situation in Río Negro, which builds on the idea that greater investments and the expansion of oil and gas production will have a multiplier effect on other sectors of the economy, as well as a positive impact on the province's finances. However, in the light of other experiences, we believe that this strategy would jeopardise a number of productive economic activities that, in the medium to long run, could form a basis for an alternative and more sustainable development model to the 'oil enclave'. (Observatorio Petrolero Sur & Enlace por la Justicia Energética y Socioambiental, 2018)

As we will see, the Mesa's work is also embedded in prior and ongoing struggles at the local, regional and global levels. Contrary to those who are pushing for the development of unconventional hydrocarbons, the Mesa believes that achieving a truly sustainable transition inevitably requires popular engagement and empowerment. Its activities are also informed by more theoretical and conceptual debates on low-carbon transitions and development. Chief among these are discussions around the just transition concept and related principles aimed at promoting and guiding a transition that minimises the labour and social impacts of the transformation or closure of unsustainable economic activities (ILO, 2015). Through its work, the Mesa expands on and deepens the just transition concept by focusing on a scenario where it is less about moving *away* from a carbon-intensive and extractivist development model, than of preventing a transition *towards* it. This means imagining and pushing a transition scenario and narrative that is more enticing and powerful than the one that is currently being offered by national and provincial authorities, and their powerful corporate allies.

THE ARGENTINA OF HYDRAULIC FRACTURING (OR FRACKING)

In this past decade, one issue has persistently dominated the energy debate in Argentina: the formation of unconventional hydrocarbons known as 'Vaca Muerta'. Located in the Neuquén basin in Northern Patagonia, the formation is estimated to be one of the largest reserves of 'technically recoverable resources' – i.e. shale gas and oil – in the world (EIA, 2013). The area of unconventional gas and oil stretches across four provinces – Neuquén, Río Negro, Mendoza and La Pampa – and over 30,000 km², an area the size of Belgium. Given their tremendous potential, the national government has drawn plans for the exploitation of these reserves through the highly contested process of hydraulic fracturing or fracking. Through the injection of a high-pressure mix of water, sand and chemicals to release gas and crude oil trapped in rocks, the fracking technique not only requires huge volumes of water but also pollutes groundwater sources, emits the potent greenhouse gas methane, and even causes small earthquakes. To date, over thirty concessions have been awarded for exploration and exploitation. While most of these are still in their infancy, two concessions, the first operated by a joint public-private consortium (YPF-Chevron)[2] in Loma Campana and the

---

2 In April 2012, the National Congress re-nationalized 51 per cent of YPF's shares in the hands of the Spanish Repsol, turning the country's main hydrocarbon extraction company into a state-run mixed capital company (see Bertinat et al., 2014; Cabrera Christiansen, 2014; Pérez Roig et al., 2016).

second by Tecpetrol – part of the Techint group – in Fortín de Piedra, have begun drilling and extracting.

While the words 'Vaca Muerta formation' historically designates a specific geologic formation, 'Vaca Muerta' is also used to refer to a larger geographic area with unconventional hydrocarbon potential in north-western Patagonia, including areas where drilling does not target the Vaca Muerta geologic formation. In other words, Vaca Muerta describes a wide range of non-conventional hydrocarbon formations, including several tight gas-drilling[3] and extraction projects extending to the provinces of Neuquén and Río Negro. When we adopt this broader definition, the number of active sites of unconventional hydrocarbon exploitation swells considerably to include six more fields, five of which are in Neuquén – El Orejano (YPF and Dow), Aguada Pichana Este (Total, YPF, Wintershall, PAEG), La Amarga Chica (YPF and Petronas), Lindero Atravesado (PAE and YPF), Centenario (Pluspetrol) – and one in Río Negro (Estación Fernández Oro [YPF]).

In addition to describing a geologic formation and list of concessions, the term 'Vaca Muerta' is also used to designate the infrastructures and processes, both upstream and downstream from exploration, drilling and extraction. These include, among others, pipelines, specific inputs for each stage of exploitation, waste treatment plants and refineries. It describes a complex and far-reaching network of existing and planned infrastructure, including roads, housing and services. In short, Vaca Muerta can be described as a megaproject that involves a wide range of local, regional, national and international public and private entities, each with their own distinctive roles and interests. Chief among these are foreign-owned corporate and financial entities – in some cases registered in a tax haven – whose priority is to maximise returns on investment and remunerate shareholders (on the role of multinationals see Snell, in this volume). As a megaproject, Vaca Muerta takes on national and even global proportions that largely surpass the geologic formation. In Argentina alone, it spans across six provinces directly: Neuquén, Río Negro, Mendoza, Chubut, Buenos Aires and Entre Ríos (see Figure 5.1; Álvarez Mullally et al., 2017).

Presented as the backbone for national and local economic development, the government's plans for a vast Vaca Muerta region further attest to Vaca Muerta's national and even global dimension (SPTIP, 2014, 2015, 2016). Among other things, government plans count on a significant increase in the regional population because of hydrocarbon

---

3 Tight gas is natural gas trapped within a rock with extremely low permeability – typically limestone or sandstone.

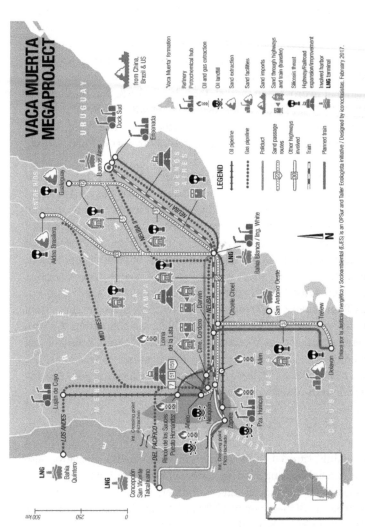

Figure 5.1 Vaca Muerta Megaproject (Source: Di Risio, 2017:16; licensed under the creative commons (CC-BY-SA): http://ejes.org.ar/VacaMuertaMegaproject.pdf (last accessed June 2019).

development. This will require new public amenities and services such as schools, hospitals and housing, as well as new or refurbished transport infrastructure. This includes plans to modernise the railway connecting Vaca Muerta to the industrial port of Bahía Blanca on the Atlantic coast, 850 km away, to transport the huge amounts of silica sands, a raw material that is essential for fracking. This massive development of infrastructure raises a number of concerns. The fact, for instance, that infrastructures are generally financed through public funds, leads to a further increase in public debt. In an unpredictable and highly volatile international hydrocarbon market, there is a real risk that the expected tax windfalls from hydrocarbon development will not cover the costs associated with new infrastructure development. The Vaca Muerta megaproject also opens the doors to highly controversial Public Private Partnerships.

The 2015 presidential elections saw the victory of a new government with an export-focused economic agenda. For the newly elected president, Mauricio Macri, Vaca Muerta is an essential means through which to address Argentina's energy and structural economic problems. Since the mid-2000s, the imports of oil and gas have further weakened an economy that was already reliant on imports of manufactured goods and is historically dependent on the export of commodities to secure a steady inflow of foreign currencies. The recently released National Energy Plan (August 2018) forecasts that, by 2027, dollar revenues from hydrocarbon exports – the majority of which will come from the Vaca Muerta megaproject – will exceed those generated through agricultural commodity exports, the country's main historical exporting sector (SGE, 2018a). In October 2018, the country began re-exporting gas following almost ten years of sustained imports.

FRACKING IN RÍO NEGRO

While mainly situated in Neuquén, the Vaca Muerta formation also extends to neighbouring Río Negro where its exploitation is expected to have major economic, social and environmental repercussions. Drawing on the expertise of the economist Mariana Fernández Massi and political scientist Diego Pérez Roig, the Mesa provided an opportunity to both take stock of the province's economic situation and to discuss the local effects of hydrocarbon development. As part of the Mesa, two smaller working groups were organised to focus on the hydrocarbon sector and the connections between economic diversification and environmental sustainability.

Río Negro is a territorially and economically diverse province that spans over 203,000 km², from the Atlantic Coast to the Andes mountain range. It comprises several 'economic regions'. These include the Andean region, with its focus on tourism, hops and fine fruits, and the *Linea Sur* (Southern Line) with its sheep and goat rearing, as well as mining activities – quarries, salt, diatomite, iron. The Atlantic zone's economy relies on fishing, tourism and port activity in the coastal town of San Antonio Oeste, and sheep and fodder production in the Lower Río Negro Valley. This is also the case for the north-east of the province. The Middle Valley (*Valle Medio)* and Upper Valley (*Alto Valle)* specialise in fruit production and horticulture. Horticultural and fruit-farming activities depend on a central irrigation system of more than 130 km and a secondary irrigation system that reaches out to the Middle Valley. The Upper Valley produces over 80 per cent of the national production of pears and apples. The Middle Valley also specialises in livestock and beekeeping. The only area with a history of oil extraction (since the 1960s) is Catriel in the north of the province. As this brief overview shows, the province's economy relies on a range of largely agriculture-related activities – and in particular fruit and meat production. The agriculture and fisheries sector has historically been a major purveyor of jobs in the province. At 20.2 per cent of total registered jobs in 2017, it was only surpassed by retail and trade (20.7 per cent) and the services sector (38.4 per cent) (OEDE-MTEySS, n.d.).

The province's decision in 2013 to develop unconventional gas exploration and extraction was justified by a worsening public deficit and rising debt. In 2016–17 alone, provincial public debt skyrocketed by an estimated 2800 per cent. Furthermore, the share of foreign currency debt in overall public debt has significantly risen in recent years (DNCFP, 2018). One priority for provincial governments across the country has been to both reduce public spending and find new sources of income. In 2006, a new amendment was added to the National Hydrocarbons Law (Law No. 17.319) transferring subsoil resource ownership to the provinces. Provincial governments were subsequently responsible for awarding hydrocarbon concessions and regulating corporate activities. In return, they collected the royalties amounting to approximately 12 per cent of the value of what is extracted (and declared) by concessionaries. As a result, oil royalties have come to represent an increasingly important source of foreign currencies for cash-strapped provinces like Río Negro.

Building on existing projects such as the Estacion Fernández Oro (EFO) tight gas field in the Upper Valley, the provincial government has subsequently favoured an 'oil rentier' approach to economic development. Stretching over 192 km^2 and at least two urban areas, specialising in pears and apples, the YPF-run EFO accounted for 2.35 per cent of national and 54.98 per cent of provincial gas production during the first ten months of 2018, and 7.33 per cent of YPF's gas production during that same period.[4] Some 160 unconventional wells are currently in production in EFO, and numbers are expected to reach the 220 mark in 2022 (*Vaca Muerta News*, 2018; SGE, 2018b). The last two years have seen an increase in the processing capacity of EFO's sorting and compression plant, as well as its transport capacity through connections to the trunk pipelines.

The provincial government has expanded the productive area well beyond the Upper Valley to areas that a priori are not connected to any of the known hydrocarbon basins. The provincial and national governments' plans include the building of new oil-related infrastructure, including a hydrocarbon waste treatment plant in Catriel, trunk connection pipelines, improvements to the Roca railway – all projects with major social and environmental implications. The expansion of the hydrocarbon frontier has led to a loss of fertile agricultural land used for the production of fruit and vegetables for the purposes of gas extraction and new real estate or infrastructure projects. From 2011 to 2016, and coinciding with the expansion of fracking, an estimated 15,000 ha of productive farm lands in Neuquén and Río Negro have been lost (Giacinti & Palmieri, 2018). According to the National Agrifood Health and Quality Service, between 2009 and 2014, the town of Allen, for example, lost 409 ha out of its 3,200 ha of productive land as a consequence of hydrocarbons (Rodil, 2015).

To justify their efforts, provincial and national authorities focus on the economic benefits, and especially the new employment opportunities and increased tax revenues for the cash-strapped province. The Mesa provided an opportunity to counter the dominant claims and to paint a very different picture. In particular, discussions highlighted that the hydrocarbon sector's relative isolation from other productive sectors in the province, even in a situation of rising investments and extraction, means that it has relatively modest effects on the provincial economy as a whole in terms of job creation. As was previously highlighted, the sector's development, through royalties and taxes, has led to a growth in state-funded infrastructures, services – public and private – and other 'unproductive' activities. This has not, however, led to the

---

4   YPF is the country's leading oil and gas extraction company. According to data from the Secretariat of Energy, in the first ten months of 2018, its total production represented 34 per cent of the country's total gas and 48 per cent of oil.

substantial development of new industries or other highly productive and job-intensive activities. Quite to the contrary, and as we have noted, the development of the hydrocarbon sector led to a decline in the agricultural sector. This in turn places further pressure on the productive performance of the 'hydrocarbon enclave', which increasingly serves as an important source of jobs and tax revenues. In this way, a dynamic and well-paid sector tends to crystallise, but whose greatest impact on employment is not that associated with hydrocarbon exploitation but, rather, through public-sector development and investment during positive economic cycles and budgetary surpluses.

The national and provincial authorities' focus on employment opportunities through regional hydrocarbon development also downplays the negative employment impacts that derive from the supplanting of more labour-intensive activities such as farming. The suppression and displacement of traditional productive activities, which were the main source of income for local communities, when combined with the ecological and health effects – access to safe drinking water, pollution – is threatening the livelihoods and balance of local rural and urban communities, and simultaneously contributing to worsen the climate crisis. This does not mean that we should simply support the status quo but rather seek to improve the practices in existence. In other words, in addition to analysing how the expansion of the hydrocarbon frontier is negatively affecting other sectors – especially agriculture – we must also look at how these other sectors can become more resilient and more economically attractive. In particular, this means looking at how agriculture, whose forms of production have also generated, in many cases, precarity and environmental degradation, can be transformed in a socially and environmentally just manner. Only by providing workers with attractive and credible employment alternatives will we be able to divert them away from the hydrocarbon sector's smoke and mirrors.

While the hydrocarbon sector's weight in terms of employment remains relatively small when compared to other economic sectors at the national level, it still constitutes an important source of livelihood for those who live in the localities where the fields and infrastructures are situated. This makes workers and communities in hydrocarbon-dependent localities particularly vulnerable to global price volatility and market disruptions. Additionally, this dependence means that local jobs and communities are especially at risk, should the decision be made – especially in the light of climate science – to phase out fossil fuels.

Despite its comparatively better working conditions and higher wages, the hydrocarbon sector, like other sectors, also faces rising job insecurity, especially following the signing by the oil workers' union, oil and gas companies, and the provincial governments of Neuquén and Río Negro of an agreement in January 2017 that worsens working conditions for workers through greater flexibility (García, 2018). Nevertheless, given their comparatively better situation, hydrocarbon-sector workers tend to be those who most actively defend and mobilise in support of hydrocarbon projects, even when if it is at the expense of their own health and safety, and that of their families and communities. There have already been five fatal work-related accidents in the Neuquén basin since the signing of the agreement. This state of affairs represents a further hurdle when attempting to imagine an alternative future.

The growth of unconventional hydrocarbons and related infrastructures and services also has major implications in terms of greenhouse gas emissions (GHG). Scientific research in the region has shown how irrigated fruit production – which is currently being threatened by hydrocarbon extraction – contributes to limiting GHG emissions and even mitigating climate change. As a group of researchers from the National University of Comahue explain:

> The rural land/urban land ratio that is required to preserve the sequestration-emission balance of greenhouse gases in the cultivation stage gives a 3/1 ratio as the optimum for this objective. This favourable balance of carbon sequestration in irrigated areas with pip fruit trees against urban emissions remained that way until 2010. The estimation of the rate of population growth for 2020 raises the need for a 10 to 20% increase of green vegetation cover in the irrigated area to preserve the balance and be a sustainable locality. (Mendía et al., 2017:1)

Beyond the loss of sink capacity due to land-use change, the large-scale exploitation of hydrocarbons is in clear contradiction with Argentina's international climate commitments, especially given fracking's greater carbon footprint when compared to more conventional methods of hydrocarbon extraction (see, for example, Sosa, 2016 or Howarth, 2016). As was previously noted, by requiring vast quantities of water and energy, the exploitation of hydrocarbons through fracking also has direct impacts on the local environment. This aspect becomes even more critical in the context of climate change. Official reports have shown how climate change has already started to affect the region and projections point to a likely worsening of water stress and desertification over the coming decades (Gobierno de Argentina, 2015). The priority, given this likely scenario, should be to implement development measures that prioritise ecosystem integrity, water-resource preservation and productive diversification as ways to improve adaptive capacities and resilience.

Through its combination of laissez-faire economics and top-down decision making, the Río Negro government is profoundly reconfiguring the province's economy. The rapid development of new wells in the Upper Valley area – particularly in the Estación Fernández Oro and Allen production areas – and concession of large areas of the Middle Valley for exploration of conventional and non-conventional hydrocarbons, was forced through without public debate or the consultation of local communities (Von Sprecher, 2018). The mainstream local and national media provided a biased and uncritical coverage highlighting its supposed benefits for the economy and in terms of energy security, and this, despite the growing opposition, especially at the local level. Through their laissez-faire approach to economic policy (what we have termed 'targeted de-plannification'), provincial and national authorities have left traditional economic sectors – that have historically provided jobs for a large portion of the population – to fend for themselves against powerful economic and financial interests (Álvarez Mullally, 2015). This was especially true for fruit production, which is increasingly dominated by a handful of powerful economic interests, and this to the detriment of small and medium producers. Similar processes are at play in the so-called *Linea Sur* where the provincial government's non-interventionist approach is facilitating the opening of several open pit mines (gold and uranium), despite local opposition.

The provincial government's growing focus on unconventional hydrocarbons and its growing dependence on oil- and gas-related incomes, will come at a high social, economic and environmental cost. Unless we are able to bring about a fundamental reorientation, the social and environmental impacts will become irreversible within the next few years. The decision to launch the Mesa largely came out of this observation and a shared desire to imagine an alternative future for Río Negro – one that draws on and strengthens existing social and community networks, improves the quality of life for the local population, and protects the local environment and climate. In particular, it builds on a long and rich history of socioenvironmental struggle in the province, and on the idea that just transition will not be possible unless it both serves the people and it is also driven by them. This can only happen if, at both the sectoral and community levels, the right conditions and mechanisms for dialogue are in place. At the sector level, truly representative bipartite and tripartite social dialogue mechanisms are needed to address labour-related issues such as working conditions, vocational training, or labour reinsertion. At the community level, it is fundamental to open wide and democratic spaces such as public consultations, deliberative assemblies and other mechanisms, especially to guarantee the participation of the most affected. Empowering frontline communities is key to allow for a more informed and inclusive transition. As we show in the following section, these can be successfully implemented by drawing on Río Negro's long and rich history of socioenvironmental struggle.

A LONG AND RICH HISTORY OF SOCIOENVIRONMENTAL STRUGGLE

For any transition project to be socially and environmentally just, it needs to be rooted in local struggles and realities. For participants in the Mesa, this means learning from and building on past and existing experiences and making sure that those who will be most directly affected by the planned hydrocarbon projects are also the most actively involved in imagining an alternative future (Blanco & Mendes, 2006). The past twenty to thirty years have been marked by a series of socioenvironmental mobilisations involving neighbourhood groups (*vecinos*) and social and environmental organisations. In many cases, they were successful in stopping various projects and coming up with credible and sustainable alternatives. Notable examples include the campaign for the remediation of land contaminated with lead in San Antonio Oeste, or the removal of PCB transformers in Sierra Grande.[5] Local acts of resistance against the Calcatreu open pit gold-mining project in the southern region not only put a halt to the project but laid the foundations for the 2005 Anti-Cyanide Law that banned the use of cyanide and mercury in mining. In 2011, the provincial legislative body repealed the law and the exploitation project is now running again.

The history of past struggles against extractive projects continues to inspire local populations and inform their outlooks and actions. In recent years, it has led to the creation of *vecino* (neighbourhood) and cross-sector assemblies where citizens can share their ideas and organise. In several localities, environmental assemblies have demonstrated an extraordinary ability to collectively strategise and coordinate their actions. A potent example is the successful mass mobilisation in 2017 to halt the construction of a nuclear power plant in Sierra Grande, at the coastal zone of the province (forming part of a package of Chinese investments in Argentina).

---

5 See, for example, the map of socioenvironmental conflicts drawn up by the Union of Education Workers of Río Negro at its first socioenvironmental congress (UnTER, 2016).

On the issue of Vaca Muerta and unconventional hydrocarbon development, local groups have organised hundreds of public talks, mobilisations, artistic and sports events throughout the province, which resulted in the banning of fracking in the municipalities of Allen, Cinco Saltos, Conesa, Coronel Belisle, Chimpay, Choele Choel, Estación Fernández Oro, Lamarque, Luis Beltrán, Pomona, Viedma and Villa Regina; and there are projects to ban fracking presented at the local legislative bodies of General Roca and Cipolletti. However, in most cases, the provincial government and the judicial system reacted by arguing that the subsoil is not owned or managed by the municipality but by the province; thus, some of these municipal ordinances were declared invalid. Catriel provides an interesting example of local, grass-roots mobilisation. Situated in the north of the province and a centre for conventional oil exploitation since the 1960s, Catriel was the site of mass popular opposition to plans to build a 300-hectare oil waste treatment plant. Going against the will of the people and despite a change in the municipal charter (*magna carta*) that precisely prohibits this type of undertaking, the provincial government – through the Secretariat for the Environment and Sustainable Development – and the municipality of Catriel allowed the plant project to go ahead and took the case to court.

In addition to its strong *vecinos* movement, Río Negro is also home to an active labour movement that has played an important role by enabling societal and environmental debates among workers in sectors directly and indirectly affected by the rapidly changing provincial circumstances. Interesting and relevant examples of trade union activism include the '*Repensar el Estado*' discussion sessions, organised by the Association of State Workers. Another example is the Chico Mendes Socioenvironmental Department and the training body of the Rodolfo Walsh School, of the Union of Education Workers of Río Negro. The authors of this chapter have been involved in these initiatives as members of the Observatorio Petrolero Sur and of the Chico Mendes Department.

The Mesa also draws on and engages with other campaigns and movements outside of Río Negro. At the national level, it engages with a nationwide anti-fracking network of local and sub-national grass-roots resistance that has succeeded, as is the case at the provincial level in Entre Ríos and in localities, in getting local governments to ban fracking (Di Risio, 2017). At the regional level, it draws on experiences in other Latin American countries. At its first public meeting, for instance, the Mesa invited Alberto Acosta, economist and former President of the Ecuadorian Constitutional Assembly, to share his experience of the Yasuní ITT project in Ecuador's Yasuní National Park. Another Ecuadorian, Maximiliano Proaño, was also invited to share his experience of the Mesa

Ciudadana de Energía para Magallanes (Magallanes Energy Transition Table, Chile) in neighbouring Chile. In May 2018, the Mesa Ciudadana launched its own 'Citizens Proposal for Energy in Magallanes' (Proaño U, 2018). Proaño also shared his knowledge of other similar initiatives in other parts of the continent.

CONCLUDING REMARKS

Having developed a shared understanding of Vaca Muerta's broader political, economic, environmental and social implications, and drawing on past and present citizens' campaigns and initiatives, the Mesa can now start constructing an alternative, socially and environmentally just development trajectory for Río Negro. It goes without saying that the elaboration of a shared diagnosis and process was far simpler than the development of a shared plan of action. Such a plan requires factoring in a wide range of variables and identifying and agreeing on deep social, cultural and political changes. It also means providing a transformational and aspirational counter-narrative that is more powerful than the jobs- and tax-centred one currently on offer by the provincial and national governments, and the corporate interests they serve.

As a first step, participants in the Mesa collectively identified four priority areas of action: organic agriculture, sustainable tourism, industrialisation of wool production, and renewable energies. A fifth economic area, non-metalliferous mining, was also singled out. For each of these areas, the plan was to review existing and related projects and initiatives, and then look at the extent to which they could be applied to the Río Negro context. Specific proposals for the hydrocarbon sector were also discussed in the Mesa. Possible avenues include the adoption of a moratorium, exclusion zones, tightening of environmental controls and taxation. Again, the Mesa agreed to analyse existing measures and similar experiences elsewhere, and on that basis, elaborate its own proposals that are attuned to provincial realities. The goal is to draft, as in Magallanes, a 'citizens proposal' for Río Negro.

While its work is still ongoing at the time of writing this chapter (April 2019), what lessons can be learnt from the Mesa experience? A first lesson relates to the importance of deconstructing the dominant corporate and government (provincial and national) narratives, and developing shared assessments based on past and lived experiences and struggles as a necessary first step to imagining an alternative future for the region – and to a certain extent the country. A second crucial lesson is that securing a just transition is as much about implementing policies and measures to protect the most vulnerable workers and communities, as it is about empowering workers and communities and creating the processes through which they can make their voices heard. In other words, the act of publicly coming together to develop a shared assessment and imagine a future without fracking is a big part of what makes the transition 'just' and what distinguishes the Mesa's transition from that of the national and provincial governments.

In the Río Negro context where socioenvironmental actors are confronted by powerful forces and a one-size-fits-all transition narrative centred on jobs and energy sovereignty, it becomes especially important to not only develop a convincing counter-narrative but to also provide an autonomous safe space in which to voice concerns and share viewpoints and aspirations. Greater efforts could undoubtedly be made to extend the Mesa and incorporate new voices, especially within frontline communities most directly impacted by hydrocarbon exploitation, as well as the productive sectors potentially affected by a possible transformation.

In short, the Mesa's ambition is to mobilise local communities and draw on their lived experiences to develop a powerful counter-narrative about possible just futures for our societies and our environment. Framing the discussion in terms of transition has provided participants with a powerful framework through which to deconstruct the extractive projects currently being promoted in the province, both by opening the door to alternative horizons (the why) and focusing attention on the importance of process (the how) when constructing and imagining other possible futures.

ACKNOWLEDGMENTS

In addition to the authors, this document is the product of over ten years of work by the Observatorio Petrolero Sur. It was also inspired by discussions with, and valuable suggestions and ideas from Joaquín Turco, María Belén Álvaro, Rodolfo Kempf, Lorena Riffo, Gustavo Lahoud, Mariana Fernández Massi, Diego Pérez Roig and those who attended the Mesa's first two meetings.

REFERENCES

Álvarez Mullally, M. 2015. *Alto Valle perforado: el petróleo y sus conflictos en las ciudades de la Patagonia Norte*. Ciudad Autónoma de Buenos Aires: Ediciones del Jinete Insomne.

——, L Arelovich, F Cabrera & D di Risio. 2017. *Informe de externalidades. Megaproyecto Vaca Muerta*. Buenos Aires: EJES – Enlace por la Justicia Energética y Socioambiental está conformado por Taller Ecologista y Observatorio Petrolero Sur.

Bertinat, P, E D'Elia, R Ochandio, M Svampa, E Viale & Opsur. 2014. *20 Mito y realidades del fracking*. Buenos Aires: El Colectivo. Colección Chico Mendes.

Blanco, DN & JM Mendes. 2006. 'Aproximaciones al análisis de los conflictos ambientales en la Patagonia. Reflexiones de historia reciente 1980–2005', *Ambiente & Sociedade*, vol. IX, no. 2, pp. 47–69.

Cabrera Christiansen, F. 2014. 'Tras la expropiación de YPF: El Estado empresario avanza sobre los yacimientos no convencionales', *Energía y Equidad*, vol. 4, no. 4, pp. 15–30.

Di Risio, D. 2017. *Vaca Muerta Megaproject. A Fracking Carbon Bomb in Patagonia*. Taller Ecologista Rosario & Observatorio Petrolero Sur in EJES (Enlace por la Justicia Energética y Socioambiental), Argentina, in cooperation with the Regional Office Cono Sur & Heinrich-Böll-Stiftung. http://ejes.org. ar/VacaMuertaMegaproject.pdf (last accessed May 2019).

DNCFP (Dirección Nacional de Coordinación Fiscal con las Provincias). 2018. *Información fiscal municipal y provincial. Base de datos.* 12 September. www2. mecon.gov.ar/hacienda/dncfp/provincial/info_presupuestaria/esq_juris_ serie_APNF.php (last accessed April 2019).

EIA (US Energy Information Administration). 2013. *Technically Recoverable Shale Oil and Shale Gas Resources: An Assessment of 137 Shale Formations in 41 Countries Outside the United States*. Washington, DC: US Department of Energy. www.eia.gov/analysis/studies/worldshalegas/archive/2013/pdf/ fullreport_2013.pdf (last accessed April 2019).

García, M. 2018. 'Otro obrero muerto en un yacimiento de YPF: Se dispara la cantidad de casos fatales en Neuquén', *El Extemo Sur*, 16 November. www. elextremosur.com/nota/otro-obrero-muerto-en-un-yacimiento-de-ypf-se-dispara-la-cantidad-de-casos-fatales-en-neuquen/ (last accessed April 2019).

Giacinti, M & A Palmieri. 2018. *Libro Blanco de la Fruticultura*. www.rionegro. gov.ar//download/archivos/00007961.pdf?1544450174 (last accessed April 2019).

Gobierno de Argentina. 2015. *Tercera Comunicación Nacional a la CMNUCC*. www.argentina.gob.ar/ambiente/sustentabilidad/cambioclimatico/ comunicacionnacional/tercera (last accessed April 2019).

Howarth, R. 2016. *Un Puente a Ningún Lugar: Emisiones de Metano y la Huella de Carbono del Gas Natural*. Mexico City: Heinrich-Böll-Stiftung. 2 January. https://mx.boell.org/es/2016/01/02/un-puente-ningun-lugar-emisiones-de-metano-y-la-huella-de-carbono-del-gas-natural (last accessed April 2019).

ILO (International Labour Organization). 2015. *Guidelines for a Just Transition Towards Environmentally Sustainable Economies and Societies for All*. Geneva: ILO.

Mendía, J, E Jockers, A Gonzalez, Z Percaz, J Forquera & M Sheridan. 2017. 'Balance del carbono en chacras regadas del Valle de Río Negro, Argentina. Primera Aproximación', Ponencia presentada en III Congreso Nacional de Ciencia y Tecnología Ambiental. Santa Fe, 31 de Julio al 3 de Agosto.

OEDE-MTEySS (Observatorio del Empleo y la Dinámica Empresarial del Ministerio de Trabajo, Empleo y Seguridad Social). n.d. *Estadísticas e Indicadores*

Regionals. www.trabajo.gob.ar/estadisticas/oede/estadisticasregionales.asp (last accessed April 2019).

Observatorio Petroloero Sur & Enlace por la Justicia Energética y Socioambiental. 2018. *Apuntes Conceptuales y Metodológicos Para el Debate.*

Pérez Roig, D, H Scandizzo & D di Risio. 2016. *Vaca Muerta: La Construcción de una Estrategia.* Ciudad Autónoma de Buenos Aires: Ediciones del Jinete Insomne.

Proaño U, M (ed.). 2018. *Propuesta Ciudadana de Energia para Magallanes.* Santiago de Chile: Heinrich-Böll-Stiftung. https://cl.boell.org/sites/default/files/paginas_propuesta_ciudadana_de_energia_para_magallanes_ok_160518.pdf (last accessed April 2019).

Rodil, D. 2015. 'Avance de la frontera hidrocarburífera sobre suelo productivo. Estación Fernández Oro, Alto Valle del Río Negro', *VII Jornadas Argentino Uruguayas de Economía Ecológica ASAUEE.* Neuquén.

SGE (Secretaría de Gobierno y Energía). 2018a. *Argentina Energy Plan – Guide Line.* www.argentina.gob.ar/sites/default/files/plan_energetico.pdf (last accessed April 2019).

——. 2018b. *Producción de Petróleo y Gas por Pozo (Capítulo IV). Base de Datos.*

Sosa, SM. 2016. 'Es el metano estúpido', *OPSUR,* 18 November. www.opsur.org.ar/blog/2016/11/18/es-el-metano-estupido/ (last accessed April 2019).

SPTIP (Subsecretaría de Planificación Territorial de la Inversión Pública de la Nación). 2016. *Estudios Estratégicos para el Desarrollo Territorial de la Región Vaca Muerta. Plan Estratégico Territorial. Avance III.*

——. 2015. *Estudios Estratégicos para el Desarrollo Territorial de la Región Vaca Muerta. Plan Estratégico Territorial. Segunda etapa. Informe Final.*

——. 2014. *Estudios Estratégicos para el Desarrollo Territorial de la Región Vaca Muerta. Primera etapa. Informe síntesis.*

UN (United Nations). 2018. *Concluding Observations on the Fourth Periodic Report of Argentina.* E/C.12/ARG/CO/4. Economic and Social Council, Committee on Economic, Social and Cultural Rights. 1 November.

UnTER (Unión de Trabajadoras y Trabajadores de la Educación de Río Negro). 2016. *Mapa de Problemáticas y Conflictos Ambientales de Río Negro.* www.unter.org.ar/node/14366 (last accessed April 2019).

Vaca Muerta News. 2018. 'Allen consolida su perfil petrolero', *Vaca Muerta News,* 28 February. http://vacamuertanews.com.ar/ver_noticia.php?id=20180225105801 (last accessed May 2019).

Von Sprecher, D. 2018. 'Inversión millonaria en Allen para una ciudad cada vez más petrolera', *Río Negro,* 25 February. www.rionegro.com.ar/allen-millonarias-inversiones-consolidan-su-perfil-petrolero-IN4492569/ (last accessed April 2019).

6
Resource rich and access poor: Securing a just transition to renewables in South Africa

Sandra van Niekerk

INTRODUCTION

Since 2008, load shedding – the controlled process of cutting electricity supply on a rotational basis – has become a common feature of everyday life in South Africa, with the most recent, and extreme instance occurring in early 2019. For much of February and March 2019, the country went through weeks of stage four load shedding – the highest level of load shedding implemented to date. Load shedding is the visible tip of a deep crisis in the South African electricity system. Indeed, load shedding appears as a luxury for the many thousands of South Africans who are off-grid or connected to the grid but too poor to afford electricity. Energy poverty – that is, the lack of adequate access to energy services or connection to the grid and inability to afford electricity – continues to form a major problem in South Africa, and this, despite significant progress following the 1994 elections. In 2013, an estimated 43 per cent of the population was considered as energy poor (SEA, 2015:51). Widespread energy poverty fits into a wider web of deep – and even deepening – social and environmental injustices in South Africa. Almost 25 years after the end of apartheid, South Africa continues to face endemic poverty, crime and corruption, and an unemployment rate of over 37 per cent (StatsSA, 2018).

High levels of energy poverty are also coupled with growing concerns about the health and environmental impacts of South Africa's existing state-run and coal-dependent energy sector. Coal provides an estimated 83 per cent of South Africa's total generating capacity (DOE, 2018a). The coal-dominated energy sector is also an important purveyor of jobs and livelihoods, especially within low-income and black majority communities. These same communities are also the most exposed to the local environmental and health risks associated with coal extraction and combustion. Despite increasingly widespread agreement on the need to phase out coal and increase the share of renewables in the overall energy mix, coal's centrality in the South African economy and society has led to sharp disagreements as to how big the share of renewables should be, and how quickly the transition should happen (compare Reitzenstein et al., in this volume). This has led to important debates, tensions and misunderstandings within and between government, trade unions and environmental groups.

Attempts have been made to bridge these differences, most notably by framing renewable energy developments as a means of not only reducing – or at least stabilising – the country's greenhouse gas emissions, but also of tackling energy poverty, stimulating economic growth and providing new 'climate jobs'. Whether or not renewables can do this will fundamentally depend on how they are introduced and developed. Hence, the need to focus on the transition process.

Debates are further complicated by the fact that the just transition issue is intertwined with debates on the desirability of public or private ownership of energy production and distribution. This issue is high on the political agenda given the fact that Eskom, the massive and vertically integrated state utility is in the midst of severe financial, governance and technical crisis. Local government, which is responsible for 40 per cent of electricity distribution, is also facing major financial, governance and capacity constraints. As we will see, successive governments have promoted a neoliberal agenda centred on energy market liberalisation and the privatisation of Eskom, arguing that this is a necessary step to ensure a stable, reliable and abundant electricity supply. In January 2019, President Ramaphosa in his State of the Nation Address, called for the unbundling of Eskom into three component parts – generation, transmission and distribution (RSA, 2019). In the face of powerful opposition from the Congress of South African Trade Unions (COSATU), the South African Federation of Trade Unions, the National Union of Mineworkers and the National Union of Metalworkers of South Africa (NUMSA), renewables are increasingly being used as a Trojan horse through which to liberalise and privatise not just renewables but the electricity sector in general.

In the popular discourse, unions' opposition to private sector-led renewables initiatives is often being framed as an opposition to renewable energy per se, and this, despite the fact that powerful unions, such as NUMSA, are committed renewable energy advocates. However, unlike the government and in line with their calls for energy democracy, they believe that only a socially owned renewable energy sector can contribute to redress past and present social injustices. As far back as 2011, they spoke of the need for a socially owned renewable energy sector. It should be noted, however, that they have been less clear on how big a portion of the energy mix renewable energy should represent. Interestingly, they are also the union that has come under most fire for their apparent opposition to renewable energy when they opposed the signing of 27 Power Purchase Agreements with renewable energy independent power producers in 2017. This subsequently forced them to clarify their position by reasserting that they are not opposed to renewables per se but rather to their privatisation (Karl Cloete, 2018a, b).

NUMSA's position echoes those of other unions and partner organisations around the world that are actively campaigning for a pro-public renewables approach. As Sean Sweeney and John Treat from Trade Unions for Energy Democracy (TUED) explain, and despite claims to the contrary, the private sector is not currently driving the transition away from fossil fuels to renewable energy, and will unlikely do so in the near future given that it is simply not profitable enough for them to do so. They and other partisans of energy democracy promote a 'not-for-profit "public goods" approach to the energy transition [which] can radically alter the prospects of renewables and allow us to effectively pursue climate and decarbonization targets' (Sweeney & Treat, 2017:6). This is the case of Public Services International (PSI), the global union federation of public-sector unions, which actively supports public energy ownership, including in renewable energies. It also argues for the re-municipalisation of energy production and distribution, a point that, as we will see, is especially relevant to the South African context (PSI, 2017).

The argument of this chapter is that debates about the low-carbon transition are not just centred on how electricity is produced but also for whom and by whom. Securing a just transition to a low-carbon energy system subsequently involves more than simply helping energy workers and their communities to transition away from coal, but also guaranteeing adequate access to clean energy for all (energy democracy). Achieving this, I argue, requires a pro-public approach that assigns a key role to Eskom, the state utility, as well as to local governments whose role and promise in this regard is examined more closely.

THE POLITICAL ECONOMY OF
ELECTRICITY IN SOUTH AFRICA

The South African economy has historically been shaped by the Minerals Energy Complex. The capitalist interests deeply embedded in this system continue to have a major influence on government actions, and more specifically on those related to electricity. This has to do with the fact that the mining industry's profit margins, as well as those of related industrial and manufacturing sectors, hinge on the availability and delivery of cheap electricity and cheap labour. As Ben Fine and Zavareh Rustomjee explain, 'the South African economy is uniquely dependent on electricity and is uniquely electricity-intensive' (1996:8). Cheap electricity has historically been possible thanks to the country's abundant coal reserves, mainly in the north of the country, which continue to account for 96 per cent of electricity generation. The result is a deeply carbon-intensive economy (Williams, 2018:240), responsible for over half of total regional emissions (Brahmbhatt et al., 2016:107). The South African energy sector accounts for approximately 84 per cent of these emissions.

Established in 1923 by the government of the Union of South Africa, Eskom (or Escom as it was then known) has historically dominated the national – and regional – electricity landscape. This vertically integrated state utility is responsible for the generation, transmission and partial distribution of electricity in the country, with local governments (as we will see below) responsible for the rest of distribution. Eskom currently generates approximately 95 per cent of South Africa's energy, mostly from coal, with 17 percent coming from gas, hydro and nuclear. At present, Eskom only plays a limited role in relation to renewable energy. Its only wind farm, the Sere Wind Farm, generates a mere 100 MW of power capacity. While the World Bank did grant Eskom a loan to build a Concentrated Solar Plant (CSP) plant at Kiwano in the Northern Cape, it ended up reallocating the funds to a battery storage project in July 2018 – officially for financial reasons (ESI Africa, 2018).

As mentioned in the introduction, energy market liberalisation and the privatisation of Eskom have come to occupy an increasingly important position on the national political agenda. The first tentative steps towards privatisation date back to the 1980s, in an economic and ideological climate marked by state disengagement, and growing economic and political pressures against the apartheid regime. Interestingly, it was with the end of apartheid and under ANC (African National Congress) rule, and in the context of a 'self-imposed' structural adjustment programme, that the first real attempts were made to privatise Eskom and liberalise the energy sector. A decisive moment was the passing of the Eskom Conversion Act in 2001 that transformed Eskom into a state-owned corporate entity required to pay dividends and taxes, and subject to commercial principles of operation (Eberhard, 2004:1). While the state remains the utility's sole shareholder, it now operates like a private company. Gentle (2009) has argued that this change of status represented a marked shift away from an energy utility that was meant to supply energy in the public interest towards a neoliberal model 'by which the electricity needs of the consumer may be satisfied in the most cost-effective manner, subject to resource constraints and the national interest' (Electricity Act of 1987, quoted in Gentle 2009:50).

Rather than making things better, the corporatisation of Eskom precipitated the utility's increasingly unsustainable and precarious financial and governance situation. With a net loss for 2017–18 of South African rand (R) 2.3 billion, R 19.6 billion in irregular expenditure and a debt of R 380 billion, Eskom is currently in an extremely precarious financial state. Its financial difficulties are the product of years of under-investment and rising operating costs, as well as allegations of large-scale corruption (Eberhard & Godinho, 2017). Its situation has worsened as a result of dropping electricity demand and the financial costs – debt servicing, massive overruns – associated with recent investments in two new coal-fired power plants – Medupi (4764 MW) and Kusile (4800 MW) – that are funded through a loan from the World Bank (2007). Eskom has to service a huge debt, which is projected to rise to R 600 billion (approximately US$ 42 billion) within the next four years. The recent drop in electricity demand is both the result of an economic downturn and a series of energy-efficiency measures put in place following two instances of forced load shedding (2008 and 2015).

Local government's role in the electricity supply chain is also being challenged. Part B of Schedule 4 of the Constitution of the Republic of South Africa assigns local government the task of distributing electricity and gas. While Eskom distributes approximately 60 per cent of electricity, it supplies only 15 per cent of customers – mostly energy-intensive mines and associated industries. Local government is responsible for distributing the remaining 40 per cent to the rest of mostly individual customers. Municipalities are therefore at the heart of the South African electricity system. In some rare instances, municipalities not only distribute electricity but generate it as well for local use. Until 2001, for example, the City of Johannesburg owned and ran the Kelvin power station in Johannesburg. This is also the case of the coal-fired Athlone power station in Cape Town (180 MW capacity), which was decommissioned in 2003. These examples notwithstanding, however, municipalities are primarily responsible for distributing electricity that they buy in bulk from Eskom. The National Energy Regulator of South Africa (NERSA) is responsible for both setting the price municipalities pay for bulk electricity, and the price they can charge consumers in their jurisdiction. The sale of electricity has historically been an important source of income for municipalities. It has often been used to cross-subsidise other, non-income generating services, including services to low-income households. It is estimated that approximately

26.8 per cent of local government revenues are derived from the selling of electricity (Montmasson-Clair et al., 2017).

As of 1997, in an attempt to restructure and consolidate electricity distribution, the government pushed for the creation of six regional electricity distributors (REDs) (Eberhard, 2007). The plan was for Eskom and municipalities to hand over all distribution assets to the newly formed entities. Given its disastrous financial implications, municipalities vehemently resisted these efforts and ultimately succeeded in shelving the proposal in 2010 (Baker & Philips, 2019). In an echo of current union efforts to resist the unbundling of Eskom, trade unions organising at the local government level, particularly the South African Municipal Workers Union (SAMWU), also opposed the restructuring plans, on the grounds that the creation of stand-alone REDs was a first step to the privatisation of electricity.

Following the 1994 elections, and as part of the transition to a democratic state, municipalities were both amalgamated and increased in size so that today 257 municipalities cover the country wall-to-wall. As a result, municipalities are now responsible for delivering service in a much larger area, for a much larger population, further increasing the financial and capacity strains on many of them (February, 2018). The amalgamation process and financial difficulties also coincide with often severe capacity and governance problems. The 2016–17 auditor-general's (AG) report on local government noted ongoing problems in the audit results of municipalities including a lack of accountability, weak internal controls and poor governance. Only 33 municipalities out of 239 that were audited received a clean audit (AG South Africa, 2018:2). In 2009, the Department of Cooperative Governance and Traditional Affairs published a report, 'State of Local Government', which identified a range of service-delivery and governance issues facing municipalities, including enormous service-delivery challenges and backlogs, poor accountability relationships with communities, corruption and fraud, poor financial management, and a lack of scarce skills (COGTA, 2009). Little has changed since the publication of this report.

One of the major and persistent challenges facing municipalities is the enormous debt that they owe Eskom. By September 2018, the level of debt reached R 17 billion. This resulted in Eskom cutting bulk electricity to some municipalities for non-payment, leaving residents and businesses without power (Paton, 2018). The debt is partly due to the non-payment of electricity bills by consumers. In some cases, certain municipalities are not transferring the money that they do receive to Eskom. While non-payment by residents is part of the reason for the huge municipal debt, it is also because many municipalities are struggling financially, with about half of all municipalities (128 out of 257) in financial distress at the end of 2016–17 (National Treasury, 2019).

Throughout the apartheid years (1948–94), electricity generation, transmission and distribution were heavily weighted in favour of big mining companies and associated industries. Through Special Pricing Agreements with Eskom, globally diversified, energy-intensive and high-emitting mining companies like BHP Billiton and Anglo-American could source electricity at exceptionally low prices, while consuming up to 44 per cent of available electricity. In contrast to the preferential access to electricity enjoyed by mining and manufacturing, most households – apart from middle-class households – have no or limited access to electricity. Before 1994, only 36 per cent of South Africans were connected to the grid (Baker et al., 2014). Lack of access to electricity for the vast majority of the poor, black working class further exacerbated poverty and inequalities in apartheid South Africa. There were significant improvements in the immediate post-apartheid years. From 36 per cent in 1994, the percentage of the population connected to the grid reached 87 per cent in 2012 (SEA, 2015:27). This was due, in part, to Eskom's ambitious plan to extend access across the country and take the grid into regions and neighbourhoods that were not previously connected. As McDonald notes, 'South Africa has electrified low-income areas on a scale, and at a pace, that is unprecedented in modern history' (McDonald, 2009:1).

With Eskom's progressive corporatisation from 1987 onwards, and as its financial situation deteriorated and supply problems worsened, however, the electricity roll-out slowed down considerably. This was accompanied by rising electricity prices. Over the last ten years alone, electricity tariffs have risen by 400 per cent. The result has been continuously high levels of energy poverty in the country, with a large number of households still either not connected or unable to afford the increasingly high tariffs that Eskom is charging, with extremely high disconnection rates as a result.

THE ELECTRICITY SYSTEM IN TRANSITION: CONTESTATIONS AROUND RENEWABLE ENERGY

As we have just seen, the South African energy system and its main utility company, Eskom, is in crisis. As Baker and colleagues have argued, a form of forced energy transition is taking place as South Africa moves from 'an era of "energy opulence" to one of restraint imposed by a series of infrastructural, economic, environmental and physical constraints' (Baker et al., 2014:793). This crisis precedes and is independent from the current concern for the climate, and the growing interest in renewables.

In South Africa, the introduction of renewable energy adds an extra layer of complexity to an already complex situation. The challenge now becomes of both addressing the energy sector's structural difficulties while simultaneously increasing the share of renewables in the energy mix. However, renewable energy is not simply the replacement of one energy source by another. It also implies moving away from the current centralised, coal-dependent energy system to a more complex, multi-dimensional, diffuse one; one that could combine both centralised and localised elements, thereby providing access to electricity for all.

For successive governments, the introduction of renewables through so-called Independent Power Producers (IPP) has served as a means of further weakening and dismantling of the state-owned Eskom and state-run energy system, and delegitimising those who defend it. Over the course of the 1990s and early 2000s, COSATU mounted a series of anti-privatisation campaigns, and this, despite its participation in the tripartite alliance with the ANC and South African Communist Party (Baker et al., 2015). Its priority was to preserve Eskom's state-owned utility status. They paid relatively little attention to the type of energy and its environmental and climate impacts. It was only when, in 2007, NERSA and Eskom discussed the introduction of renewables into the energy mix that unions began to actively get involved in the renewable energy debate. For unions, the main concern was of seeing how much renewable energy should be introduced, how fast this should happen, who should produce it, and most importantly, what this would mean for workers – especially those in the coal sector. The period was also characterised by growing union involvement in the international climate debate and engagement with the just transition concept, especially within the International Trade Union Confederation (see Rosemberg, in this volume).

Following an initial and inconclusive attempt at introducing renewable energy through a feed-in tariff ('REFIT' scheme, 2009), NERSA launched a public procurement programme in March 2011, the year in which South Africa was to host the UNFCCC Climate Conference (COP17) in Durban. The Renewable Energy Independent Power Producers Procurement Programme's (REIPPPP) stated purpose is to mobilise private capital in support of the South African renewable-energy sector through a series of bid windows whereby a given amount of MW in specific technology categories (mainly solar and wind) are made available and opened to bidding. The winning IPPs sign 20-year Power Purchase Agreements (PPAs) with Eskom that guarantee them a certain income stream. As a state entity, Eskom was barred from placing bids.

The REIPPPP was intended as an instrument through which South Africa could both reduce its greenhouse gas emissions (GHG) and simultaneously increase overall electricity supply. It should be noted that under the programme, renewable energy would still remain a relatively small component of the overall energy mix. The 2010 Integrated Resource Plan (IRP) (which was updated in 2011) capped newly built renewable energy at 42 per cent of the overall energy mix, which amounts to a little over 9 per cent of total electrical energy production by 2030. Nuclear, coal, gas and energy-efficiency measures, and not renewables, remained the core elements of the overall energy mix (DOE, 2011). In a 2015 report, the Department of Energy praised the programme, noting that it '[was] now receiving worldwide acclaim' and that, in 2014, South Africa was the tenth largest solar market in the world for installations above 5 MW (DOE, 2015:10). While frequently touted as a major success story, a growing body of research is pointing to the programme's limitations, especially in terms of its implementation (see, for instance, Baker & Wlokas, 2015; McDaid, 2014; Mthembi, 2015).

Government efforts through the REIPPPP to increase the share of renewable energy were coupled with a commitment to secure a just transition for workers – especially in the coal sector. In its Intended Nationally Determined Contribution (INDC), in the lead-up to the Paris Climate Conference (COP21), the South African government refers to the need for a 'just transition to a climate resilient economy and society, taking into account local and indigenous knowledge, gender considerations, as well as social, economic and environmental implications' (Government of South Africa, 2015:2). The government's seeming commitment to renewables and a just transition have not always translated into concrete and decisive actions. Coal continues to form a central part of the energy mix and while the draft 2018 IRP aims to bring coal down to less than 20 per cent of the total energy mix in 2050, it also licences a short-term increase in coal consumption – adding 1000 MW of coal over the 2023–24 period (DOE, 2018b). It is also worth noting that South Africa's industrial policy remains centred on the heavily carbon-emitting mining sector (Baker et al., 2015:8). This gap between discourse and reality has sent mixed signals about the government's true intentions, and its commitment to securing a low-carbon energy transition that is both beneficial to workers and energy consumers.

In response to the government's actions, South African unions sought to educate the public on climate change, and especially its negative social impacts – such as rising food and electricity prices, and water shortages. They combined these efforts with calls for a transformative, public ownership-centred energy policy that simultaneously addressed the climate crisis and secured energy democracy. In 2010, this even led unions to work together with environmental justice groups around climate change and the just transition. These interactions gave birth, in 2011, to the One Million Climate Jobs Campaign whose ambition, as its title suggests, was to create at least a million decent, people-driven jobs that reduce greenhouse gas emissions, build capacity to adapt to the impacts of climate change, and provide vital public services such as water, energy and sanitation (OMCJC, 2011).

At around the same time, COSATU adopted a forward-thinking and progressive policy on climate change that included a call for a just transition away from a high carbon, coal- and oil-dependent economy to a low-carbon economy powered through renewables (COSATU, 2011). For COSATU, securing a just transition was vital to garner worker support for a low-carbon transition. It also noted that the development of new industries as part of the transition towards renewable energy must not act as an excuse for lowering wages and social benefits. Instead, the COSATU calls for 'decent jobs that are paid at living wages, that meet standards of health and safety, that promote gender equity, and that are secure' (COSATU, 2011:56). It is also worth noting that the union associates the just transition concept with the need to shift towards a 'fundamentally transformed society' (ibid.).

COSATU was not the only South African union to seriously engage in the climate and renewable energy debate and adopt the just transition concept. In 2012, the powerful NUMSA, then an affiliate of COSATU, recognised that 'at the centre of the climate change question is how to rapidly move away from fossil fuels towards renewable energy' (NUMSA, 2012:10). In particular, it called for the creation of a socially owned renewable energy sector composed of a 'mix of energy parastatals, cooperatives, municipal-owned entities and other forms of community energy enterprises' (ibid.).

This growing trade union involvement with the climate issue was short lived. By 2014, the focus of COSATU and its affiliates (including NUMSA) had shifted away from issues such as climate change, and towards internal tensions, which culminated in the expulsion of NUMSA from COSATU in November 2015. As with the government, this produced a growing gap between the movement's discourse and the reality of its actions and positions. The earlier efforts to engage with the climate issue were not followed through and little concrete content was given by the unions to the notion of a just transition, or a socially- owned renewable energy sector and what this would practically imply.

The fragility and shallowness of trade union approaches to renewables was particularly visible during the crisis that opposed Eskom and the

REIPPPP in 2016–17. In March 2017 – and despite the fact that these plants had already been scheduled for decommissioning over the next few years – Eskom threatened to close down various coal-powered generation plants on the grounds that it could no longer keep them running and at the same time cover the costs related to bringing renewable energy IPPs on line. Buying into Eskom's arguments, energy and mining unions accused IPPs of being indirectly responsible for the loss of jobs in mining and energy. Their mobilisation conveniently served Eskom's interests, since it had been refusing, since July 2016, to sign 27 PPAs with renewable energy IPPs that were due to come on line.

Trade union involvement in the conflict between Eskom and REIPPPP projected a distorted image of what unions stood for, ultimately strengthening the hand of the government and those who support market-based solutions to the climate and energy crises. It presents them as being against renewables when in fact they have consistently opposed *privatised* renewables. This position was reasserted by NUMSA when it stated that

> ... the mandate of Renewable Energy projects must be to achieve service provision, meet universal needs, decommodify energy and provide an equitable dividend to communities and workers directly involved in production and consumption of energy [with] a socially owned RE sector ... governed via democratic control through constituency-based governing councils tasked with achieving a strict social mandate. (Karl Cloete, 2018a)

Despite unions' commitment to publicly owned renewables, many people interpreted their opposition to IPPs as a blanket rejection of renewable energy, rather than just of privatised renewable energy, leading to tensions between trade unions and the environmental justice groups. This was the case in 2018, when NUMSA took the government to court in an effort to stop them from signing the 27 PPAs (see above). Their move provoked reactions of outrage among environmental organisations, such as Greenpeace, who saw this as a deliberate attack against renewables and an attempt to defend the coal industry (Greenpeace, 2018). NUMSA was quick to clarify that it was opposed to the privatisation of renewable energy and not an outright opposition to renewables. They have since then reiterated on various occasions their support for publicly owned renewable energy (Karl Cloete, 2018a, b; Jim, 2018; NUMSA, 2019).

BREAKING THE PUBLIC RENEWABLES CONUNDRUM?

Given the myriad financial and governance problems that currently plague the South African energy system, and the current deadlock between unions and the government, the possibility of both turning Eskom into a leader in renewable energies and retaining its vertically integrated utility status is going to be extremely challenging. With the future of Eskom dominating the energy debate, far less attention has been paid to local governments and the role they can play in relation to renewables and the just transition. And yet, local governments deserve our attention for two main reasons. First, and as was previously noted, municipalities, in their constitutionally sanctioned function as electricity distributors, already occupy a central position in the South African energy debate. Secondly, renewable energy, as an energy source, lends itself well to a decentralised and locally managed system of generation and distribution. Thus, getting Eskom and local governments to combine their efforts to roll out renewable energy appears as a potentially viable and effective way of increasing access to clean, affordable electricity. The idea is not to prevent community organisations, cooperatives and other forms of small-scale ownership from playing their part in renewable energy development but rather of inserting these smaller entities into a well-coordinated and publicly run energy sector so as to ensure equitable, reliable and affordable access to clean energy. Eskom and local governments would subsequently be tasked with overseeing and running the process.

A recent study identified a number of different roles that local governments could play as part of a transition (Montmasson-Clair et al., 2017). These include building generation capacity, procuring energy either from embedded generation or from IPPs, and playing a facilitating role in promoting community renewable energy projects, cooperatives and so on. In addition to this, municipalities can and already are, playing a role in increasing energy efficiency.

While recognising that 'embracing the transition is no longer a choice but a necessity if our energy sector is to survive' (SALGA, 2018:3), local governments are relatively isolated (and on the frontlines) when it comes to dealing with the day-to-day challenges and dysfunctionalities of the current energy system. In the case of renewables, they are often left to grapple with and adapt to an unclear regulatory framework and unstable financial environment. Faced with load shedding, increasing tariffs, non-payment, informal connections to the electricity grid, wealthier households and businesses able to go off-grid as they install PV panels on their roofs, and the impact that this has on their revenue stream, municipalities have had to take initiatives and develop their own approach to renewable energy. These challenges exacerbate already existing inequalities between local governments, as the more financially sound and wealthy municipalities are in a stronger position to deal with these challenges, particularly the large metropolitan municipalities ('metros'). Unless there is greater regulatory clarity and certainty, the danger is therefore of widening the gap between the wealthier municipalities that have a richer resource base to draw on and are thus able to take initiatives on their own, and poorer municipalities that have to grapple with high levels of unemployment and poverty, inadequate service delivery, and weak institutional capacity.

The City of Cape Town is one 'metro' that has taken important steps to develop locally sourced and distributed renewables. In 2018, it initiated a Constitutional Court bid to win the right for municipalities to buy power directly from independent sources and not only from Eskom (Kim Cloete, 2018). The case has yet to be resolved, but should Cape Town win, it could put them at loggerheads with unions which, as we have seen, oppose privatised renewable energy. An alternative could be for municipalities to develop their own generation capacity. To date, this has only been attempted on a small scale. Examples include landfill gas extraction to electricity plants (for instance in eThekwini), sewerage biogas to energy plants (for example in Johannesburg), and small-scale hydro power plants (for example in the !Kheis Municipality in the Northern Cape) (SEA, 2015:39). Such facilities generally employ a limited number of people. Such projects also present a number of challenges. These include, among others, finding the funds to finance them, lack of clarity on whether an electricity generation licence is required from NERSA, and limited support from the Department of Energy.

A further challenge for municipalities relates to the rapid increase, since 2016, in the number of small-scale embedded generation (SSEG) units, and this, despite the lack of a clear regulatory framework or tariff structure. A growing number of municipalities actually authorise SSEG installations on their electricity grid, and have a tariff system in place to compensate households and businesses who feedback surplus electricity into the grid (SALGA, 2017). One major problem with SSEG is that it can potentially exacerbate existing inequalities. The wealthier sections of the population can afford to install solar PV panels and become largely independent from the grid. According to estimates, for municipalities this can contribute to a drop in their electricity revenues by 17–25 per cent in residential areas (Baker & Philips, 2019). In 2017, and in an attempt to better regulate SSEG, amendments were finally made to the Electricity Regulation Act exempting power generation facilities of less than 1 MW from a generation licence, even though they are still expected to register with NERSA. Municipalities have also begun to implement measures of their own. In Cape Town, for instance, all SSEG systems need to be officially registered.

Energy efficiency offers perhaps the greatest potential for municipalities in terms of job creation – a key issue in relation to just transition. A core advantage of these jobs is that they can easily be created in every municipality. Examples of energy-efficiency work include retrofitting solar water heaters, solar water heater maintenance, and retrofitting residential and commercial buildings with a range of energy-efficient measures such as lighting and ceilings. In the early 1990s and as part of the government's initial attempts to respond to the major housing shortage, thousands of so-called 'Reconstruction and Development' (RDP) houses were built, but without ceilings – to bring down costs and increase turnover. Over the last decade, many municipalities, such as the City of Cape Town, have embarked on projects to retrofit ceilings into these houses to make them more energy efficient.

Trade unions in the local government sector have, for a long time, recognised the need to engage in the local energy debate. The Independent Municipal and Allied Trade Union (IMATU), for instance, 'strongly believes that organized labour must be part of the immediate strategies to address electricity generation challenges as well as the broader energy concerns regarding transmission and distribution' (IMATU, 2015). The SAMWU has also been very active, developing and pushing policy proposals relating to service delivery in the electricity sector. With the mooted establishment of the REDs in the 2000s, SAMWU took an active part in the debate. Unfortunately, as with other South African unions, internal conflicts within SAMWU stretching back to 2014 have prevented it from focusing on broader strategic and policy-related issues in the local government sector.

There is increasing interest from poor and working-class communities in urban areas, as well as from rural communities who do not have access to electricity, for the development of local initiatives in the form of cooperatives that can generate renewable energy. However, these community initiatives face a number of major hurdles. These include, among others, access to land, finances, and a regulatory framework for feed in tariffs. Local government could play a facilitating role in relation to these issues. They could also create the conditions for communities within their jurisdictions to set up small-scale renewable energy projects.

CONCLUSION

South Africa provides an original and interesting case for a just transition. Its originality not only stems from the fact that it is a highly carbon-intensive and coal-dependent economy (see chapters by Goods and Snell in this volume), but also from the fact that large sections of the population still cannot afford, or simply do not have adequate access to electricity. As a result, the challenge for stakeholders involved in discussions on a low-carbon transition becomes not only of securing a just transition for coal workers and coal communities, but also of providing adequate, equitable and affordable access to electricity for all.

While both the government and trade unions frequently express their commitment to renewables and refer to the just transition concept, their conflicting understandings of how and who should drive the transition have led to a form of paralysis. While successive governments have promoted a liberalised energy market and the privatisation of electricity generation and distribution, trade unions argue that only a state-run and state-owned electricity system can both secure a low-carbon future and energy democracy. The risk of a deregulated electricity system, unions argue, is of worsening an already unjust system in which the rich benefit from privately owned, clean and abundant electricity, whereas the – majority black – poor are left without proper access to electricity and are exposed to the environmental and health risks of an underfunded and coal-dominated state-owned electricity system.

As we have seen, local governments, given their already existing roles in energy distribution, could potentially form the basis for a low-carbon and people-centred electricity system. Given the fact that renewables are a diffuse energy source, municipalities could both generate more electricity themselves, and in this way increase access to electricity, but could also play a key role in ensuring that everyone within their jurisdiction has equitable access to renewable energy.

However, for this to happen will require breaking the current deadlock and a radical shift in the existing power dynamics around electricity in South Africa. It will also require a concerted effort to strengthen Eskom and local government institutionally and financially. Given the current state of Eskom and local government, and the powerful influence of coal interests on the country's economic and political system, the transition to a new and more socially, environmentally and politically just electricity system will only happen through a united front of trade unions, environmental groups, women's groups, youth groups and frontline community organisations. That is perhaps the real challenge facing just transition in South Africa – the ability of all these organisations to unite around a common vision of a people's energy system.

REFERENCES

AG South Africa. 2018. *Consolidated General Report on the Local Government Audit Outcomes: MFMA 2016–17*. AG-SA. www.agsa.co.za/Portals/0/Reports/

MFMA/201617/GR/MFMA2016-17_FullReport.pdf (last accessed April 2019).

Baker, L, J Burton, C Godinho & H Trollip. 2015. *The Political Economy of Decarbonisation: Exploring the Dynamics of South Africa's Electricity Sector*. Cape Town: Energy Research Centre, University of Cape Town.

——, P Newell & J Phillips. 2014. 'The political economy of energy transitions: The case of South Africa', *New Political Economy*, vol. 19, no. 6, pp. 791–818.

—— & J Phillips. 2019. 'Tensions in the transition: the politics of electricity distribution in South Africa', *Environment and Planning C: Politics and Space*, vol. 37, no. 1, pp. 177–196.

—— & HL Wlokas. 2015. *South Africa's Renewable Energy Procurement: A New Frontier?* Cape Town: Energy Research Centre, University of Cape Town.

Brahmbhatt, M, R Bishop, X Zhao, A Lemma, I Granoff, N Godfrey & DW te Velde. 2016. *Africa's New Climate Economy: Economic Transformation and Social and Environmental Change*. London & Washington, DC: New Climate Economy & Overseas Development Institute. https://newclimateeconomy. report/workingpapers/wp-content/uploads/sites/5/2016/11/Africa_ NCE_2016_final_1.pdf (last accessed April 2019).

Cloete, Karl. 2018a. 'Op-ed: NUMSA supports a transition from dirty energy to clean renewable energy', *Daily Maverick*, 15 March. www.dailymaverick.co.za/ article/2018-03-15-op-ed-numsa-supports-a-transition-from-dirty-energy-to-clean-renewable-energy/ (last accessed April 2019).

——. 2018b. 'Coal jobs, independent power producers (IPPs) and energy sovereignty: A NUMSA perspective', *South Africa Labour Bulletin*, vol. 42, no. 3, pp. 4–6.

Cloete, Kim. 2018. 'Cape Town pushes ahead with plans for more renewable energy', *Creamer Media's Engineering News*, 26 July. www.engineeringnews. co.za/article/cape-town-pushes-ahead-with-plans-for-more-renewable-energy-2018-07-26 (last accessed April 2019).

COGTA (Department of Cooperative Government and Traditional Affairs). 2009. *State of Local Government in South Africa: Overview Report. National State of Local Government Assessments*. Working Documents. Department of Cooperative Government and Traditional Affairs.

COSATU (Congress of South African Trade Unions). 2011. *A Just Transition to a Low-carbon and Climate Resilient Economy: COSATU Policy on Climate Change: A Call to Action*. COSATU.

DOE (Department of Energy). 2018a. *2018 South African Energy Sector Report*. November 2018. Pretoria: DOE.

——. 2018b. *Draft Integrated Resource Plan 2018*. August 2018. Pretoria: DOE.

——. 2015. *State of Renewable Energy in South Africa*. Pretoria: DOE.

——. 2011. *Integrated Resource Plan for Electricity 2010–2030*. Pretoria: DOE.

Eberhard, A. 2007. 'The political economy of power sector reform in South Africa', in D Victor & TC Heller (eds), *The Political Economy of Power Sector Reform*, pp. 215–253. Cambridge: Cambridge University Press.

——. 2004. 'The political economy of power sector reform in South Africa', Working Paper WP-06 Programme on Energy and Sustainable

Development, Stanford University. www.gsb.uct.ac.za/files/Stanford
PSREberhardSep2004final.pdf (last accessed April 2019).

—— & C Godinho. 2017. *Eskom Inquiry Reference Book: A Resource for
Parliament's Public Enterprises Inquiry, Civil Society, Journalists and Engaged
Citizens.* University of Cape Town Graduate School of Business. www.
gsb.uct.ac.za/files/Eskom%20Enquiry%20Booklet%20Sept%202017.pdf (last
accessed April 2019).

ESI Africa. 2018. 'Eskom ditches CSP project for battery storage development',
ESI Africa's Power Journal, 27 July. www.esi-africa.com/eskom-ditches-csp-
project-for-battery-storage-development/ (last accessed April 2019).

February, J. 2018. 'Local government in South Africa: Mostly corrupt, largely
dysfunctional', *Daily Maverick,* 26 April. www.dailymaverick.co.za/opinionista/
2018-04-26-local-government-in-south-africa-mostly-corrupt-largely-
dysfunctional/ (last accessed April 2019).

Fine, B & Z Rustomjee. 1996. *The Political Economy of South Africa: From
Minerals-energy Complex to Industrialization.* London: Hurst & Company.

Gentle, L. 2009. 'Escom to Eskom: From racial Keynesian capitalism to neoliberalism (1910–1994)', in D McDonald (ed.), *Electric Capitalism: Recolonising
Africa on the Power Grid,* pp. 50–72. Cape Town: HSRC Press.

Government of South Africa. 2015. *South Africa's Intended Nationally Determined
Contribution.* 25 September 2015. www4.unfccc.int/sites/submissions/
INDC/Published%20Documents/South%20Africa/1/South%20Africa.pdf
(last accessed April 2019).

Greenpeace. 2018. 'Sabotaging renewable energy will continue to cost South
Africans', Press release, 15 March. www.greenpeace.org/archive-africa/en/
Press-Centre-Hub/Sabotaging-renewable-energy-will-continue-to-cost-
South-Africans/ (last accessed April 2019).

IMATU (Independent Municipal and Allied Trade Union). 2015. 'IMATU
believes intervention in electricity supply is crucial', *IMATU News,* 5 June.
www.imatu.co.za/imatu-believes-intervention-in-electricity-supply-is-
critical/ (last accessed April 2019).

Jim, I. 2018. 'Eskom's IPP project fails test for a just transition', *Politicsweb,* 30
July. www.politicsweb.co.za/news-and-analysis/the-ipp-project-fails-the-test-
for-a-just-transiti (last accessed April 2019).

McDaid, L. 2014. *Renewable Energy Independent Power Producer Programme
Review 2014.* Electricity Governance Initiative of South Africa. http://
thegreenconnection.org.za/doaction/wp-content/uploads/2015/04/EGI-
REI4P-review-2014.pdf (last accessed April 2019).

McDonald, D. 2009. 'Electric capitalism: Conceptualising electricity and capital
accumulation in (South) Africa', in D McDonald (ed.), *Electric Capitalism:
Recolonising Africa on the Power Grid,* pp. 1–49. Cape Town: HSRC Press.

Montmasson-Clair, G, K Kritzinger, L Scholtz & M Gulati. 2017. *New Roles
for South African Municipalities in Renewable Energy – A Review of Business
Models.* Discussion Paper. South African-German Energy Partnership. TIPS.
www.tips.org.za/research-archive/sustainable-growth/green-economy-3/

item/download/1396_002e3326e878ae6b82bb4272925b31eb (last accessed April 2019).

Mthembi, F. 2015. 'Lost in procurement: An assessment of the development impact of the Renewable Energy Procurement Programme', in L Mytelka, V Msiman & R Perrot (eds), *Earth Wind and Fire: Unpacking the Political, Economic and Security Implications of Discourse on the Green Economy*, pp. 111–144. Johannesburg: Mapungubwe Institute for Strategic Reflection.

National Treasury. 2019. *Budget Review 2019*. Pretoria: Republic of South Africa. www.treasury.gov.za/documents/national%20budget/2019/review/FullBR. pdf (last accessed April 2019).

NUMSA (National Union of Metalworkers of South Africa). 2019. NUMSA Press Conference Statement on Eskom, 21 February. www.numsa.org.za/article/ numsa-press-conference-statement-on-eskom/ (last accessed April 2019).

——. 2012. 'Building a socially-owned renewable energy sector in South Africa', in *NUMSA 9th National Congress: Final Economic Resolutions*. www.numsa. org.za/wp-content/uploads/2012/07/00042_final_economic_resolutions_ v2_240612_-_.pdf (last accessed April 2019).

OMCJC (One Million Climate Jobs Campaign). 2011. *One Million Climate Jobs: A Just Transition to a Low Carbon Economy to Combat Unemployment and Climate Change*. Cape Town: OMCJC, Mowbray.

Paton, C. 2018. 'Municipal debt to Eskom balloons to R17bn', *Business Day*, 14 November. www.businesslive.co.za/bd/companies/energy/2018-11-14- municipal-debt-to-eskom-balloons-to-r17bn/ (last accessed April 2019).

PSI (Public Services International). 2017. *Resolutions Adopted by Congress Volume 1: People Over Profit*. congress.world-psi.org/wp-content/uploads/2017/12/ EN-Vol-1-Draft-PoA-adopted-by-Congress-Nov-2017.pdf (last accessed April 2019).

RSA (Republic of South Africa). 2019. *President Cyril Ramaphosa: 2019 State of the Nation*. 7 February. www.gov.za/speeches/president-cyril-ramaphosa- 2019-state-nation-address-7-feb-2019-0000 (last accessed April 2019).

SALGA (South African Local Government Association). 2018. *Energy Summit Declaration*. 2018 Energy Summit 'Defining the energy future of local government', Johannesburg, Gauteng, 7–9 March 2018. www.salga.org.za/ Documents/Documents%20and%20Publications/Documents/Final-Energy- Summit-Declaration-2018.pdf (last accessed April 2019).

——. 2017. *Status of Small Scale Embedded Generation (SSEG) in South African Municipalities*. Pretoria: SALGA. www.salga.org.za/SALGA%20Energy%20 Summit%202018/Energy%20Summit%20Web/Document/Status%20of%20 Small%20Scale%20Embedded%20Generation.pdf (last accessed April 2019).

StatsSA (Statistics South Africa). 2018. *Quarterly Labour Force Survey. Quarter 3: 2018*. Pretoria: StatsSA. www.statssa.gov.za/publications/P0211/ P02113rdQuarter2018.pdf (last accessed April 2019).

SEA (Sustainable Energy Africa). 2015. *State of Energy in South African Cities*. Westlake: SEA. www.cityenergy.org.za/uploads/resource_322.pdf (last accessed April 2019).

Sweeney, S & J Treat. 2017. 'Preparing a public pathway: Confronting the investment crisis in renewable energy', Working Paper No. 10. New York: Trade Unions for Energy Democracy. http://unionsforenergydemocracy.org/wp-content/uploads/2017/10/TUED-Working-Paper-10.pdf (last accessed April 2019).

Williams, M. 2018. 'Energy, labour and democracy in South Africa', in V Satgar (ed.), *The Climate Crisis: South African and Global Democratic Eco-Socialist Alternatives*, pp. 231–251. Johannesburg: Wits University Press.

7

The story of coal in Germany: A model for just transition in Europe?

Alexander Reitzenstein, Sabrina Schulz & Felix Heilmann

INTRODUCTION

Through its *Energiewende*, Germany is globally recognised as a leader in the low-carbon energy transition. Rooted in the anti-nuclear movement of the 1980s and 1990s, and systemically rolled out since the early 2000s, it has succeeded in significantly increasing the share of renewables in the energy mix of a major industrial nation and in the process contributed to a rapid decline in the price of renewables. These successes notwithstanding, Germany also has a long history of coal mining. In 2018, the burning of hard coal and lignite still accounted for approximately 35 per cent of Germany's electricity mix. The coal sector's economic power has shrunk quite dramatically since the late 1950s in western Germany, and since reunification in eastern Germany. In parallel, coal-related employment went from more than 750,000 workers in 1957 to significantly less than 30,000 today. Despite its decreasing economic importance, coal, as we have just noted, continues to form an important component of the German energy system. Furthermore, for the former hard coal mining regions in western Germany as well as the lignite mining regions in Lusatia, the Rhineland and Central Germany, coal continues to shape local cultures and identities, harking back to a glorious industrial past. Associated with this is the memory of prior attempts to transition away from coal a memory that fuels a sustained opposition within the coal regions to transition efforts.

These observations point to a central contradiction in the German energy transition, what has been dubbed the 'coal conundrum' (Jungjohann & Morris, 2014). While the energy transition has succeeded in supporting the growth of renewables (especially by driving down prices), it has fallen short when it comes to eliminating coal, the fossil fuel with the highest CO_2 emissions. As a result, coal has become the main reason for Germany's failure, and by a significant margin (8 per cent), to reach its 2020 goal of reducing greenhouse gas (GHG) emissions by 40 per cent, compared to 1990 levels (see Bundesministerium für Umwelt, Naturschutz und nukleare Sicherheit, 2018a). Given the urgent need for decisive climate action, there is mounting pressure on German policy makers to commit to and carry out a rapid coal phase-out. These pressures include international commitments (especially those contained in the Paris Climate Agreement), stricter EU regulation on air pollution, increased campaigning by non-state actors (social movements, environmental groups), greater public awareness of the environmental and health-related dangers of coal, and economic pressure, including through higher CO_2 prices and investor activism.

In 2018, this eventually led the newly elected federal government to launch a 'Commission for Growth, Structural Change, and Employment', commonly referred to as the 'Coal Commission'. The commission builds on and prolongs a long history of 'consensus democracy' and multi-stakeholder deliberative processes. Its mandate was to agree on an action plan that would enable Germany to meet its domestic emission-reduction targets for 2030. In particular, this meant agreeing on a transition plan for coal that was both speedy – given the urgency of the climate crisis – and socially just for coal-dependent workers and communities. As we will show, finding the right balance between these two exigencies would prove a very challenging task, and would ultimately lead to a disappointing result from a climate standpoint, bringing into question the pertinence of existing deliberative mechanisms to address climate change. In particular, the under-representation – and even absence – of newer social and economic actors would prove an important barrier to rapid transformative change, which, from a climate perspective – and given the scale and speed at which change needs to happen – is fundamental to set the German economy on a low-carbon trajectory.

'SOCIAL DIALOGUE' AND 'DELIBERATIVE DEMOCRACY'

To understand prior and ongoing coal transition dynamics, we first need to identify and highlight some key features of the German political system. First, we should note that 'social dialogue' or 'social partnership', involving government representatives, employer groups and trade unions, has traditionally played a crucial role in Germany's economic and industrial policy. 'Social dialogue' usually takes the form of close and decentralised structures of contracts and mutual agreements between employers' associations, trade unions and workers' councils, which ensure regular, output-oriented interactions in individual industries (Dustmann et al., 2014). Through this 'autonomous bargaining system', employer associations and trade unions are responsible, for instance, for approving collective wage agreements at the sector level. In times of economic and social crisis or change, however, a system of 'informal tripartism' sets in, that includes governments in the negotiations (Lesch & Vogel, 2017:131–145). Social partnership is an important part of a corporatist approach to industrial development, especially in traditional sectors such as coal, with its high rates of unionisation (Heeg, 2012).

Secondly, given the federal nature of the German political system, policy making can be a highly complex process. Each level of the political system – the federal government, the states *(Länder)* and municipalities – has a set of clearly defined decision-making and implementation prerogatives. In the area of energy policy, most direct legislation is in the purview of the federal government. However, *Länder* have the possibility of blocking and amending laws in the *Bundesrat*, the federal council that represents the 16 federal states *(Länder)*. They can also specify and toughen regulations, make public investment decisions, and enact planning laws (cf. Appunn, 2016). An added layer of complexity sets in when federal decisions directly affect the social, economic and other policy areas of the *Länder*. In these cases, they have the possibility of either accelerating or blocking change. Thus, regional economic and social prerogatives strongly influence the national policy agenda.

The faith in 'social dialogue' and the federal political system feed into a third key element of the German political system that has and continues to shape the coal transition debate. Germany has strong 'consensus democracy' elements, whereby competing interests are organised and addressed through various multi-stakeholder processes. While it has historically proven effective when it comes to accommodating different expectations and demands, it has also tended to favour incumbent stakeholders – including industry groups, unions, established NGOs and academics – allowing them to both dictate the direction and pace of the deliberative process. As we will see, this has had major implications for discussions on coal.

A HISTORY OF GERMAN COAL TRANSITIONS

Historically, coal mining and combustion have played a vital role in fuelling Germany's economy. In West Germany, the post-war 'economic miracle' was fuelled by coal. In the 1950s, West Germany's, mostly unionised, coal- sector workers were national symbols of post-war industrialisation, and contributed to forge regional and worker identities. It was also in the 1950s that coal's use went into steady decline with the development of oil-burning and nuclear power plants. This triggered a long and arduous process of successive pit closures. Government authorities, industry and the powerful coal unions worked closely to mitigate the economic and social impacts of closures for workers and their communities, but with mixed results. The long-planned closure of the last underground mine for hard coal in late 2018 marked the end of an important chapter in Germany's industrial history. From 2019 onwards, the remaining hard coal plants would only be fuelled by imported coal. Going forward, the future of lignite mining regions, in Lusatia, Central Germany and the Rhineland, is therefore set to dominate the political debate, despite the fact that Germany still operates 25 GW of hard coal-generation capacities based on the largest imports of hard coal in the EU, including from Russia, the US, Australia and Colombia (Verein der Kohlenimporteure, 2018).

In the southern Ruhr area, hard coal mining dates back to medieval times. The first underground mines were established in the nineteenth century. Following the end of the Second World War and the division of Germany, the majority of German production of underground hard coal was situated in western Germany. Hard coal mining, together with the associated heavy industry – particularly in the steel sector – is widely regarded as the backbone of the 'economic miracle' (*Wirtschaftswunder*) that enabled Germany's rapid recovery after the war. Furthermore, the sector's economic importance when combined with the hard and dangerous work in the underground mines contributed to shape a strong sense of identity and pride among workers and within communities (Prinz & Pegels, 2018).

With financial backing from the federal government, the period stretching up to 1958 was characterised by a steady increase in coal production in West Germany. The end of the Suez Crisis, when combined with the European Coal and Steel Community's efforts to liberalise the energy sector in the mid-1950s, resulted in a major influx of oil and cheap coal. This triggered a profound crisis of the West German coal sector, which was marked by a series of colliery closures (*Zechensterben*). At the peak of hard coal production in 1957, over 600,000 workers were employed in the sector. Less than ten years later, 320,000 workers had lost their jobs (see Figure 7.1). This profound sectoral crisis did not put an end to West German coal production. On the grounds that coal was strategic to preserving Germany's energy independence, and driven by a political desire to prevent structural disruptions in the affected regions, the hard coal industry continued to benefit from substantial government subsidies. Between 1950 and 2008, hard coal mining was subsidised with €289–331 billion (Forum Ökologisch-Soziale Marktwirtschaft, 2010). A

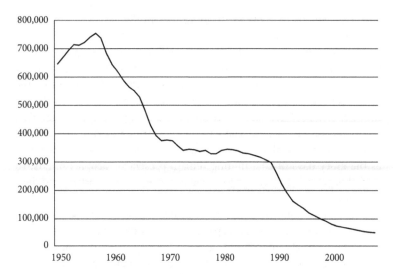

Figure 7.1   German coal employment, 1950–2008 (Source: authors' illustration based on data from Statistik der Kohlenwirtschaft e.V., 2018, 2019).

coalition formed by the coal industry, trade unions and politicians, particularly from the ranks of the Social Democratic Party (SPD), played an influential role in pushing for continued support for German coal.

While the first years of this decline in the late 1950s coincided with the final years of the 'economic miracle', allowing most workers to transfer into other jobs mainly in the metal industry (Nonn, 2001), the overall deterioration of the economy after 1962 worsened the situation and triggered a series of policy interventions. In the most affected region, the Ruhr, the second half of the 1960s was characterised by a series of structural policies. In an attempt at reviving traditional industries in the region, billions of Deutschmarks (DM) were invested in social, education and infrastructure support (Prognos AG, 2015). Following the oil price shocks of the 1970s, a further DM 2 billion were invested to modernise coal mines and plants, as well as the steel industry (Prognos AG, 2015). With a budget of DM 10 billion, the 'Action Programme Ruhr' (1980–84) was intended to revive the regional economy through measures to improve technology transfer between universities and businesses, and to increase the role of the service sector. However, the effectiveness of this and other early programmes was limited due to the initial focus on traditional industries and lack of diversification efforts. In 1986, protests in the coal regions caused policy makers to turn to more regionally focused and dialogue-based approaches to structural policy. A potent example of this was the 1987 '*Zukunftsinitiative Montanregion*' which allowed coal and steel regions to apply for funding specifically aimed at innovation measures (Hesse et al., 1991).

In the German Democratic Republic (GDR), coal also played a major role for economic development. Prior to reunification, the coal sector of Lusatia alone employed over 100,000 workers. Lignite from the open-pit mines in Lusatia and Saxony-Anhalt was essential for the GDR's industries. With German reunification, and the opening-up of the East German economy, a great many heavy industries and large parts of the coal sector were unable to compete and were forced to close down. Just as the glory days of coal mining remain present in people's minds, so does this period of crisis and its social implications for workers and their communities.

In the 1990s, following German reunification, structural policies were aimed at reducing regional disparities. In this context, EU funds took on an increasingly important role in addressing structural change, and the transition out of hard coal mining in Germany was characterised by close cooperation between the coal industry, trade unions and the relevant regional governments in western Germany. Although several attempts to rescue the hard coal sector and to shape regional economies ultimately resulted in failure, the political will to support and compensate workers prevailed. Not all measures had the desired effects. Yet this approach would still set a precedent by emphasising the need to consider the social aspects of transitions away from coal.

Given its localised and open-pit nature, the mining and combustion of lignite is generally cheaper and more profitable than hard coal. The political and economic difficulties that surrounded the phasing-out of hard coal, when combined to lignite's comparative economic advantages, contributed to delaying any real attempt at phasing out lignite. There was, in fact, a broad consensus on the need to support domestic lignite production for energy security purposes (Renn & Marshall, 2016). Historically, the lignite industry has been especially important in East Germany where 90 per cent of post-war primary energy consumption involved the burning of lignite, making it, by far, the largest global producer. The lignite sector was also closely connected to domestic heavy industry, and the country's overall economic and industrial performance largely depended on it (Kahlert, 1988).

As in West Germany, the East German lignite industry benefitted from active government support, with annual subsidies of up to DM 2.3 billion over the course of the 1980s, the equivalent of almost one-quarter of the GDR's total industrial investment (Kahlert, 1988). Following Germany's reunification in 1990, the East German lignite industry's situation changed practically overnight, as its productivity was far below western German standards. The sector struggled to compete in a liber-alised, increasingly privatised, market economy because its productivity, measured in tonnes per worker, was three times lower than that of its western German counterpart (Herpich et al., 2018). Problems arising from outdated processes and low efficiency affected not only the lignite sector but also the heavy industry of the former GDR, leading to the collapse of its industrial core and thereby reduction in energy demand. In the lignite sector, the adjustment of processes to higher standards and the integration into a market economy caused a drastic consolidation, which led to the immediate closure of many power plants and mines and a radical decline in employment, dropping from more than 112,000 workers in 1990 to 26,000 in 1995 and just above 10,000 in 2000 (Statistik der Kohlenwirtschaft e.V., 2018).

Structural policy responses during the post-reunification years were not exclusively targeted at the lignite sector but at the economy in general, with the federal government undertaking a series of measures to manage industrial decline in the East. Given the lack of established structures governing the relations between workers, industry and government, it was unable to replicate the 'managed decline' approach adopted for hard coal in the Ruhr (for a more detailed account, see Schulz & Schwartz-kopff, 2016). The federal government did however provide extensive early retirement packages to all employees over the age of 55 (Buchholz, 2008). It also negotiated severance packages with companies, including short- and part-time work arrangements. In doing so, an estimated 360,000 workers were able to gradually transition out of their old jobs and into newly created – usually low-paid and precarious – ones. The federal government also implemented various re-skilling measures to facilitate workers' transition to new jobs. This led to the retraining of 400,000 workers in the first half of the 1990s. The measures were espe-cially successful for engineers and administrative staff but far less so in the case of lower-skilled workers.

As the privatisation of lignite mines and power plants continued, the government established a state-owned company, the Lausitzer und Mitteldeutsche Bergbau-Verwaltungsgesellschaft (LMBV), which was tasked with recultivating former mining areas and cleaning up environ-mental damages. Financed through the federal budget (75 per cent) and affected *Länder* (25 per cent), and employing some 20,000 people, the LMBV initiative was, from the outset, intended as a measure to cushion the impact of the lignite industry's decline, providing new job opportu-nities, including for workers from the old industry sector.

What this brief account shows is that current debates on the phase-out of coal do not charter entirely new territory. Historical experiences and efforts to manage the coal decline inform the current transition debate, and help us to understand ongoing concerns about the potentially negative economic and social impacts of a coal phase-out. This is especially true for East Germany where the transition process proved far less effective than in the West. As we will see in the following section, a growing sense of urgency around climate change produced new pressures on the federal and state governments to take decisive action and phase out the last remaining coal mines and plants. As our account of the Coal Commission will show, the legacy of prior transition efforts, when combined with the urgent need to act on climate change, also complicated the task of developing and delivering a phase-out strategy that was acceptable to all stakeholders, in the process, highlighting the limits of the German 'consensus democracy' approach.

CLIMATE POLICY AND GERMANY'S COAL TRANSITION TODAY

Germany's status as leader in the global energy transition is increasingly being contested given its failure to meet domestic and EU energy and climate targets, and reluctance to push for more ambition. The government's reluctance to phase out coal has significantly undermined Germany's standing in the European and international climate debate (Littlecott et al., 2018). The country continues to be the largest producer of coal-related GHG emissions in Europe (see Europe Beyond Coal, 2018), and is home to seven of the continent's ten most polluting coal plants (Jones, 2016). Coal is one of the main reasons why Germany will in all likelihood fail to meet its 2020 climate targets and struggle to meet its 2030 ones (Bundesministerium für Umwelt, Naturschutz und nukleare Sicherheit, 2018a; compare Figure 7.2). Achieving the domestic 2030 climate targets would require Germany to dramatically reduce its energy-related emissions from the current 318 Mt CO_2eq to below 170 Mt CO_2eq in 2030 (Bundesministerium für Umwelt, Naturschutz und nukleare Sicherheit, 2018b). Complying with the goals of the Paris Agreement would even require a complete phase-out of coal by 2030, and a more rapid growth of renewables and related infrastructure (Climate Analytics, 2018; Umweltbundesamt, 2018).

Over the past two decades, growing public pressure and acknowledgment within national and international policy and activist circles of the need to phase out coal – especially to preserve chances of staying on a 1.5°C to 2°C trajectory (as laid out in the Paris Agreement) – triggered a series of initial steps towards decommissioning remaining older, more polluting coal plants. This in turn produced a reaction from trade unions and the coal industry, which successfully secured financial compensa-

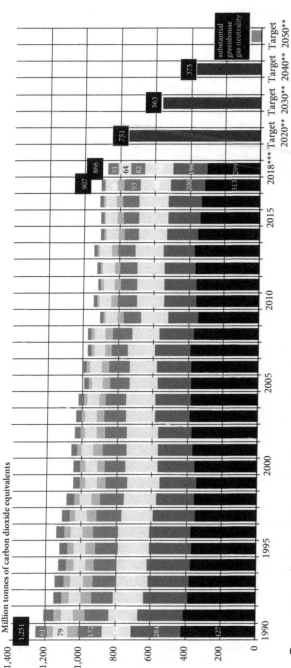

Million tonnes of carbon dioxide equivalents

Energy Industry ■ Industry* ■ Transport ■ Households ■ Commercial/Institutional ■ Agriculture ■ Waste and Waste Water ■ Other Emissions*

Emissions by UN reporting category, without land use, land use change and forestry. * Industry: Energy and process-related emissions from industry (1.A.2 & 2). Other Emissions: Other combustion (rest of CRF 1.A.4, 1.A.5 military) & fugitive emissions from fuels (1.B). ** Targets 2020 to 2050: Energy Concept of the German Federal Government (2010) *** Short-term forecast for 2018, emissions from commerce, trade & services contained in Other Emissions

Source: German Environment Agency, National Inventory Reports for the German Greenhouse Gas Inventory 1990 to 2017 (as of 01/2019) and estimate for 2018 from UBA Press Release 09/2019 (corrected)

Figure 7.2 Emission of greenhouse gases covered by the UNFCC (Source: German Environment Agency (*Umweltbundesamt*), National Inventory Reports for the German Greenhouse Gas Inventory 1990–2017, as of January 2019).

tions for the planned shutdowns. Over the 2013–17 period, for instance, €13.8 billion were earmarked to support the transition in Germany's remaining coal regions (see Deutscher Bundestag, 2018a). Through the so-called 'lignite reserve' (2016), €1.6 billion of public funding was assigned to plant operators for withdrawing 2.7 GW of capacity ahead of final closure while keeping them on standby for emergencies (Amelang et al., 2016).[1] Interestingly, the lignite reserve's standby capacity has not yet been tapped into due to the continuous overproduction of electricity in Germany (Deutscher Bundestag, 2018b), which is another indicator of the political nature of the deal.

Not surprisingly, the growing debate over climate and the need for a transition to 100 per cent renewable energy, placed coal centre stage during the 2017 federal elections. In fact, disagreements over the pace of a future coal phase-out were among the reasons behind the Christian Democratic Union/Christian Social Union (CDU/CSU), Green and Liberal Party's (FDP) failure to agree on a coalition programme. The Green Party has always been a strong proponent of an ambitious coal phase-out and elevated the topic's political importance by making it one of the key demands in the coalition negotiations. The market-liberal FDP was opposed to ambitious domestic climate action despite its support for international regimes and market mechanisms for achieving climate targets. Having failed to agree with the FDP and Greens, the CDU/CSU opened discussions with the SPD to establish a second Grand Coalition.

The coal issue also occupied an important part of these discussions and highlighted the divisions within and between Germany's two main parties – CDU/CSU[2] and SPD – over coal. The conservative CDU/CSU's internal divisions reflected concerns by influential party figures – especially at the state level – that a coal phase-out would lead to rising energy prices for energy-intensive industries. The SPD, historically well established in West German coal regions, is also divided on the issue. Throughout much of the work of the commission, the party, fearing that it would lose voters to the far-right populist Alternative for Germany (AfD), called for a slow and gradual phase-out (ZEIT ONLINE, 2018a). This was contrary to the wishes of the SPD-affiliated Environment Minister, Svenja Schulze, and various SPD parliamentarians who stressed the importance of Germany remaining a 'pioneer' and frontrunner in a

1 The alternative would have been a 'climate levy' that would have penalised older and particularly the most CO_2-intensive lignite plants.

2 CDU/CSU is actually an alliance of two parties, the Christian Democratic Union and the Christian Social Union in Bavaria. They are described here as one as they form a common parliamentary group in the Bundestag.

socially just low-carbon transition (Schulze, 2018; see e.g. Metzner et al., 2018).

Disagreements within and between the parties of the Grand Coalition favoured a less proactive approach towards ambitious climate policy. It also explains the Grand Coalition's decision to launch a commission dedicated to negotiating a coal phase-out (Reitzenstein & Schulz, 2018). The Commission on Growth, Structural Change, and Employment (or Coal Commission) was agreed on in March 2018 as part of the Grand Coalition's agreement (see also Reitzenstein, 2018). Its mission was to produce a concrete plan by late 2018 to: a) reduce the gap towards meeting the 2020 emissions reduction target to the extent possible, b) meet the domestic 2030 targets for the energy sector, c) gradually reduce and end power production from coal, including agreeing on a phase-out date and accompanying legal, structural, economic, and social measures, and d) ensure financial support for the transition and structural adaptation in affected regions.

Even though the Coal Commission process was triggered by climate policy concerns, economic pressure on coal is high and questions of profitability proved important in the discussion of transition measures such as compensation payments for the coal industry. While suggested by the Coal Commission, the German Parliament's research service, for example, found that such payments were not necessary (Deutscher Bundestag, 2018c). It should be noted that the Coal Commission's recommendations would not be legally binding on the government.

The launch of a multi-stakeholder commission was in no way exceptional (see above). However, the German government's decision to adopt a 'deliberative democracy' approach reflects its reluctance to take on the sole responsibility for the phasing-out of coal, especially given internal party debates, the mixed legacy of previous transition efforts and coal's symbolic/cultural importance in certain *Länder*. The commission was formally coordinated by the Federal Ministry for Economic Affairs and Energy, with support from the environmental, labour and interior ministries. It was made up of 28 voting members, and representatives of the three Grand Coalition parties and lignite-mining *Länder* as non-voting members. The voting members came from a diverse set of backgrounds and interest groups: representatives from coal regions, trade unions, businesses and industry, environmental NGOs, and academia/science (Bundesministerium für Wirtschaft und Energie, 2018). Given the nature of the stakeholders and the complexity of the issue, the commission was widely expected to produce a compromise position that would be too weak with respect to the urgency of the climate crisis but too ambitious for those representing the coal regions' interests (see e.g. Buchsbaum, 2018).

While not directly sitting on the commission, affected *Länder* governments were especially influential within the commission. Governments from the East German *Länder* mainly affected by the coal phase-out – Saxony-Anhalt, Saxony, Brandenburg – made sure that their voices were heard, particularly their desire for a long transition period, significant federal transfer payments, and compensation mechanisms. The government of North Rhine Westphalia, while also calling for a gradual transition, voiced its concerns about potential fall-outs, especially rising prices for its energy-intensive industries and supply insecurity.

As we have seen, business associations also traditionally play an important role in the German policy-making process. Through the presence of five of the main business associations, the business community was well represented in the Coal Commission, and could hope to influence its proposals, particularly with respect to keeping the costs for the energy-intensive industry to a minimum. But there were also differences within and between business associations. This was the case of the Federation of German Industries (BDI) whose publication of a widely recognised study on potential pathways to decarbonisation by 2050, for example, was met with harsh criticism from within its own membership (see BDI, 2018). While actively involved in CO_2 pricing discussions, some members of the German Association of Energy and Water Industries (BDEW) opposed an accelerated coal phase-out (BDEW, 2018). These positions are in sharp contrast to new business associations representing low-carbon industries, such as the German Renewable Energy Federation (BEE).

Like business, the trade union movement also provides a varied and somewhat divided landscape. The federal umbrella organisation for German trade unions, the DGB, has historically demonstrated a strong commitment to climate action and just transition at the national and supranational (EU, international) levels. Given the corporatist nature of social dialogue in Germany, however, sector-based union federations exert a higher level of power and influence and in the case of coal, they contributed to toning down the DGB's overall position in relation to the transition. Representing highly skilled workers in both local utilities and large coal power plants, the United Services Trade Union (ver.di), for example, calls for financial support to those affected by the transition from coal while also expressing its commitment to existing domestic and international climate targets. Ver.di's position diverges from that of the Union for Mining, Chemical Industry and Energy (IG BCE) which represents coal-mine workers and actively pushes for a slow coal phase-out, capitalising on its ability to mobilise its members (Kreutzfeldt, 2018). IG BCE's effective campaigning in the commission focused on slowing down the coal phase-out and ensuring significant public transfer payments for the transition, particularly to ensure a decent future for workers in coal industries.

Other important stakeholders in the commission included environmental NGOs, local civil society groups and members of the academic and scientific community. When compared to other issue areas, there was a high degree of alignment between civil society, including environmental NGOs and local activist groups, and scientific actors. Three environmental NGOs and two representatives from local communities affected by resettlements were members of the Coal Commission. Environmental groups have played an increasingly influential role in the debate and actively push for an ambitious coal phase-out by 2030. They initiated and developed innovative campaigns to pressure the commission from the outside and shape the broader energy and climate debate. As we will show below, the conflict over the Hambach Forest provides a potent example of this.

The academic and scientific community has traditionally played an important role in German decision making. Scientists are frequently consulted and scientific boards are regularly set up and convened to advise policy makers on a wide range of issues. Given their role in the climate debate, climate scientists and climate policy experts were well represented in the Coal Commission. An environmental economist was one of the four co-chairs of the commission, and five eminent scientists were voting members of the commission. Additionally, scientists were regularly invited to share their thoughts on various dimensions of the coal transition (for example, on carbon pricing). The commission's work was also informed by scientific studies and publications.

While the Coal Commission brought together and represented a wide range of stakeholders with highly diverse views on coal and climate policy, it is worth mentioning that several important groups were left out. Youth groups, representatives from countries most affected by climate change, renewable energy cooperatives and faith groups, for instance, were not represented, and would most probably have advocated for an ambitious and speedy coal phase-out. Despite its limitations in terms of composition, the commission was still perceived as an innovative effort to secure a just transition to a low-carbon economy; and one that could potentially be replicated in other industrialised contexts. In the following section, we explore the functioning of the commission as well as its internal dynamics, before analysing its outcomes.

THE COAL COMMISSION AND
THE FUTURE OF COAL IN GERMANY

A first step for the commission was to come up with a common work programme and framework conditions, as well as a shared understanding of the structural policy challenges that lay ahead. This was the focus of the first meetings of the commission. A 'sherpa' was designated to support each of the commission members, preparing the meetings and coordinating the work. Experts and academics provided input on a range of topics, including structural policy challenges in the affected regions, carbon pricing, CO_2 budgets, and the impacts of a coal phase-out on energy supply and security. The commission's work was split into two work streams. The first focused on structural change – social and economic measures to support the transition towards a low-carbon economy – and the second on developing a coal phase-out plan and agreeing on an exit date.

Throughout the process, non-voting members and other stakeholders – including federal government ministries, representatives from state governments and members of Parliament, unions and NGOs – sought to weigh in on the commission's work. At the government level, the Environment Ministry, for instance, called on the commission to respect climate policy commitments. The Ministry for Economic Affairs and Energy, on the other hand, called for the safeguarding of jobs in the coal industry as well as among suppliers and in dependent service industries (RBB, 2018). The Labour Ministry, in turn, voiced its own views on how best to achieve the structural transition (Haerder, 2018). Governments of lignite-mining states also repeatedly called for financial support towards the affected businesses, workers and communities. They were often backed by unions like the IG BCE which did not hesitate to stage protests to raise awareness on the need for a gradual and socially fair transition (ZEIT ONLINE, 2018b).

Most notably, in November 2018, just days before the expected release of the commission's final report, the Prime Ministers from the three East German coal-mining *Länder* called for a stronger focus on concrete, measurable and implementable transition measures ahead of any phase-out decision (BUND NRW, 2018). The Federal Chancellery responded by postponing the report's release, to provide the commission with more time to rework its recommendations. This was the first and only time Angela Merkel publicly intervened in the debate. External political interventions such as this were widely criticised by civil society and some media outlets, on the grounds that they undermined the commission's legitimacy and delayed the report's release (originally planned for the UN Climate Conference in Katowice in December 2018).

Discussions in the commission were also shaped by coal- and climate-related protests and mobilisations. These included an important environmental protest against plans by the German energy utility giant, RWE, to clear the 12,000-year-old Hambach Forest near Cologne to unearth more coal. The protest mobilised a range of environmental groups and drew public attention towards the commission's work and climate debate more broadly (Dobush, 2018). Opinion surveys showed that public opinion was strongly in favour of an accelerated coal phase-out and stopping the clearance of the forest (ZEIT ONLINE, 2018c). Not surprisingly the fate of Hambach became a point of discussion in the commission, with environmental groups threatening to walk out unless a moratorium was agreed upon. They ultimately succeeded in getting the final report to mention that the commission found it desirable to safeguard the Hambach Forest. During the last weeks of the commission, the international students' movement 'Fridays for Future' protesting for more ambitious climate action also played an influential role.

Following months of heated discussions, the commission published its recommendations in the form of a report in January 2019. In it, the commission calls for a complete coal phase-out by 2038. Among other things, it recommends 12 GW of early closures by 2022, €40 billion of investments in transition measures in affected regions over the period leading up to 2038, compensations for utilities in the case of early closures and for businesses and consumers in the case of rising electricity prices. The commission paid particular attention to the needs of lignite-mining regions, calling for measures to redirect their economic models to new, 'clean' industries and sectors (Bundesministerium für Wirtschaft und Energie, 2019).

Planned transition policies – outlined in the final report – are primarily targeted at improving the economic competitiveness of affected regions through investments in innovative projects as well as low-carbon technologies, and by moving state bodies/institutions to these regions. To improve the attractiveness to investors, the report suggests investing in transport and digital infrastructure. It also recommends strengthening regional research and development (R&D) institutions, especially in the area of energy transition, tax-funded R&D subsidies, and the creation of pilot economic zones to attract investments. The report discusses measures to support the transition of coal workers to new jobs as well as their early retirement, to facilitate the involvement of local civil society groups in improving social cohesion, as well as strategies to attract investors and start-ups. In an annex, and drawing on extensive evaluations of the social and economic opportunities and challenges in each of the regions, the report outlines a long list of concrete structural change projects that could be implemented immediately or in the long term, in each of the affected regions.

While the amount of funding for transition and compensation measures remained politically contentious, ultimately leading to the postponement of the report's release, there was widespread consensus on what type of transition measures were needed. All members agreed that social and economic support schemes were essential to ensure social cohesion and economic competitiveness for each of the affected regions. Where the commission members disagreed was with respect to the pace of the transition, specifically the road map and timetable – including the final exit date – for the phase-out of coal. As we have already shown, a number of regional and interest groups – unions, businesses, regional governments – were worried that a rapid transition would have detrimental short-term effects on the economy, workers and communities. Proponents of ambitious climate action, on the other hand, argued that only a rapid phase-out of coal could ensure that Germany complied with its commitments under the Paris Agreement. As could be expected from the multi-stakeholder set up, the outcome was a compromise position that was out of touch with the urgency of the climate crisis (Reitzenstein & Popp, 2019).

This difficulty in reaching a compromise in the commission reflects a fundamental challenge for climate policy makers involved in multi-stakeholder fora that are tasked with devising just transition pathways (see also Reitzenstein & Popp, 2019). The Coal Commission was in the strong tradition of negotiated, equity-based political solutions to complex political problems in Germany. In decisions of societal importance, Germany has often chosen to set up multi-stakeholder commissions, bringing all key interests to the table so as to reach a balanced decision. Considering this tradition, the outcome of the Coal Commission was an expected compromise between the main interest groups – affected states and communities, industry, unions and civil society.

A MODEL TO FOLLOW?

Reconciling climate policy requirements and the concerns of businesses, workers and communities affected by a coal phase-out is a challenging task. As we have seen, historically, the chosen approach to solving complex problems in Germany has been to convene multi-stakeholder committees to come up with consensus positions. But, as our analysis of the Coal Commission has also shown, the adoption of such an approach raises a series of fundamental questions. The first relates to the underlying motives for convening such committees. In this case, a combination of external pressures to act on climate change, and divisions between and within the parties that make up the second Grand Coalition, triggered the decision to set up the Coal Commission. In other words, rather than signalling a true commitment to deliberative democracy, the convening of the commission can also be interpreted as a convenient pretext for the German government to shy away from its responsibility to act on coal – and climate – by outsourcing the problem to others.

Secondly, questions relating to the appropriateness and legitimacy of such initiatives arise when we look at the commission's mandate, organisation and composition. Indeed, by privileging incumbent and established stakeholders over new ones, initiatives such as these favour 'lowest common denominator' outcomes. As we have found, rather than bringing about an ambitious and transformative plan, the make -up of the commission strengthened the hand of incumbent stakeholders that had the most to lose from a speedy phase-out of coal. This raises important questions about the relevance of traditional consensus solutions in the age of climate change, and the need to open deliberations to new actors. While drawing lessons from past transition efforts is important, it is also essential to recognise that these efforts were predicated on an outdated industrial and development model that is out of touch with the scale and urgency of the climate crisis.

The commission's plan is not ambitious enough for Germany to meet international expectations and commitments made in the Paris Agreement. A just transition away from coal in line with the requirements of the Paris Agreement would be of utmost relevance far beyond the German borders. In Europe and internationally, observers are watching closely to learn how an industrial powerhouse with a highly efficient industrial sector can remain competitive without relying on fossil fuels and nuclear energy for its energy supply. Delivering a successful transition to a low-carbon economy that benefits the affected regions and provides new opportunities for coal workers who are losing their jobs could inspire similar approaches elsewhere in the world. Only if Germany succeeds in raising its level of ambition and speeding up the transition out of coal in a socially and economically fair manner, can the Coal Commission serve as a blueprint for just transitions elsewhere.

ACKNOWLEDGMENT

The authors would like to acknowledge funding by the European Commission's LIFE programme which has greatly benefitted their research and work on which the chapter is based.

REFERENCES

Amelang, S, K Appunn, S Egenter & J Wettengel. 2016. 'Climate levy – the debate and proposals for cutting CO_2 emissions', *Clean Energy Wire*, 27 May. www.cleanenergywire.org/news/climate-levy-debate-and-proposals-cutting-co2-emissions (last accessed April 2019).

Appunn, K. 2016. 'German federalism: In 16 states of mind over the Energiewende', *Clean Energy Wire*, 10 March. www.cleanenergywire.org/factsheets/german-federalism-16-states-mind-over-energiewende (last accessed April 2019).

BDEW (Bundesverband der Energie- und Wasserwirtschaft). 2018. 'Positionspapier: CO_2-Bepreisung'. www.bdew.de/energie/positionspapier-co2-bepreisung/ (last accessed April 2019).

BDI (Bundesverband der deutschen Industrie). 2018. 'Climate protection needs massive investment drive according to new BDI study'. https://english.bdi.eu/article/news/climate-protection-needs-massive-investment-drive-according-to-new-bdi-study/ (last accessed April 2019).

Buchholz, S. 2008. *Die Flexibilisierung des Erwerbsverlaufs. Eine Analyse von Einstiegs- und Ausstiegsprozessen in Ost- und Westdeutschland*. Wiesbaden: VS Verlag für Sozialwissenschaften.

Buchsbaum, LM. 2018. 'Mixed mandate: Germany's new coal commission struggles to balance environment and jobs', *Energy Transition. The Global Energiewende*, 11 June. https://energytransition.org/2018/06/mixed-mandate-germanys-new-coal-commission-struggles-to-balance-environment-and-jobs/ (last accessed April 2019).

BUND NRW. 2018. 'Ministerpräsidenten torpedieren zügigen Abschluss der Kohle-Kommission', *BUND Landesverband Nordrhein-Westfalen*, 22 November. www.bund-nrw.de/meldungen/detail/news/ministerpraesidenten-torpedieren-zuegigen-abschluss-der-kohle-kommission/ (last accessed April 2019).

Bundesministerium für Umwelt, Naturschutz und nukleare Sicherheit. 2018a. 'Klimaschutzbericht 2018 zum Aktionsprogramm Klimaschutz 2020 der Bundesregierung'. www.bmu.de/fileadmin/Daten_BMU/Download_PDF/Klimaschutz/klimaschutzbericht_2018_bf.pdf (last accessed April 2019).

——. 2018b. 'Climate action in figures: Sector targets 2030'. www.bmu.de/fileadmin/Daten_BMU/Download_PDF/Klimaschutz/klimaschutz_in_zahlen_sektorenziele2030_en_bf.pdf (last accessed April 2019).

Bundesministerium für Wirtschaft und Energie. 2019. 'Kommission "Wachstum, Strukturwandel und Beschäftigung" – Abschlussbericht'. www.kommission-wsb.de/WSB/Redaktion/DE/Downloads/abschlussbericht-kommission-wachstum-strukturwandel-und-beschaeftigung.pdf?__blob=publicationFile&v=4 (last accessed April 2019).

——. 2018. 'Bundesregierung setzt Kommission 'Wachstum, Strukturwandel und Beschäftigung' ein', *Pressemitteilung*, 6 June. www.bmwi.de/Redaktion/DE/Pressemitteilungen/2018/20180606-bundeskabinett-setzt-kommission-wachstum-strukturwandel-und-beschaeftigung-ein.html (last accessed April 2019).

Climate Analytics. 2018. 'Science based coal phase-out pathway for Germany in line with the Paris Agreement 1.5°C warming limit: Opportunities and benefits of an accelerated energy transition'. https://climateanalytics.org/media/germany_coalphaseout_report_climateanalytics_final.pdf (last accessed April 2019).

Deutscher Bundestag. 2018a. 'Antwort der Bundesregierung auf die Kleine Anfrage der Fraktion BÜNDNIS90/DIE GRÜNEN – Drucksache 19/4792 – Strukturwandel in den Braunkohleregionen'.

——. 2018b. 'Antwort der Bundesregierung auf die Kleine Anfrage der Fraktion BÜNDNIS90/DIE GRÜNEN – Drucksacke 19/686 – Verfügbarkeit der Kohlekraftwerke in der Sicherheitsbereitschaft (Kohlereserve)'.

——. 2018c. 'Ausarbeitung: Stilllegung von Kohlekraftwerken', Wissenschaftliche Dienste.

Dobush, G. 2018. 'RWE getting tired of activists' Hambach Forest occupation', Handelsblatt, 13 September. www.handelsblatt.com/today/politics/coal-clash-rwe-getting-tired-of-activists-hambach-forest-occupation/23583310.html (last accessed April 2019).

Dustmann, C, B Fitzenberger, U Schönerg & A Spitz-Oener. 2014. 'From sick man of Europe to economic superstar: Germany's resurgent economy', Journal of Economic Perspectives, vol. 28, no. 1, pp. 167–188.

Europe Beyond Coal. 2018. 'Data'. https://beyond-coal.eu/data/ (last accessed April 2019).

Forum Ökologisch-Soziale Marktwirtschaft. 2010. 'Staatliche Förderungen der Stein- und Braunkohle im Zeitraum 1950–2008'. Berlin.

Haerder, M. 2018. 'Arbeitsminister Heil schlägt Sechs-Punkte-Plan für Strukturwandel vor', WirtschaftsWoche, 2 August. www.wiwo.de/politik/deutschland/braunkohleausstieg-arbeitsminister-heil-schlaegt-sechs-punkte-plan-fuer-strukturwandel-vor/22874170.html (last accessed April 2019).

Heeg, S. 2012. 'The erosion of corporatism? The rescaling of industrial relations in Germany', European Urban and Regional Studies, vol. 21, no. 2, pp. 146–160.

Herpich, P, H Brauers & P-Y Oei. 2018. 'An historical case study on previous coal transitions in Germany', IDDRI & Climate Strategies.

Hesse, J, A Benz & A Benz. 1991. Regionalisierte Wirtschaftspolitik: Das Beispiel der Zukunftsinitiative Montanregionen, Schriften zur Innenpolitik und zur kommunalen Wissenschaft und Praxis. Baden-Baden: Nomos.

Jones, D. 2016. 'Top 10 European polluters still dominated by German lignite', Sandbag, 1 April. https://sandbag.org.uk/2016/04/01/top-10-european-polluters-still-dominated-by-german-lignite/ (last accessed April 2019).

Jungjohann, A & C Morris. 2014. Energy Democracy: Germany's Energiewende to Renewables. London: Palgrave Macmillan.

Kahlert, J. 1988. Die Energiepolitik der DDR – Mängelverwaltung zwischen Kernkraft und Braunkohle. Bonn: Verlag Neue Gesellschaft GmbH.

Kreutzfeldt, M. 2018. 'RWE hilft bei Gewerkschaftsdemo', Die Tageszeitung, 23 October. www.taz.de/!5541067/ (last accessed April 2019).

Lesch, H & S Vogel. 2017. 'Working together: Germany's response to the global economic and financial crisis', in I Guardiancich & O Molina (eds), *Talking through the Crisis. Social Dialogue and Industrial Relations Trends in Selected EU Countries*, pp. 131–145. Geneva: ILO.

Littlecott, C, L Burrows, A Reitzenstein, P de Pous & R Popp. 2018. *G7 Coal Scorecard – Fourth Edition: Decision Time for Coal in Germany*. E3G – Third Generation Environmentalism.

Metzner, T, H Monath & S Ehlerding. 2018. 'Ministerpräsident Woidke: "Ein schneller Kohleausstieg stärkt die AfD"', *Der Tagesspiegel*, 1 September. www.tagesspiegel.de/politik/ministerpraesident-woidke-ein-schneller-kohleausstieg-staerkt-die-afd/22984460.html (last accessed April 2019).

Nonn, C. 2001. 'Die Ruhrbergbaukrise: Entindustrialisierung und Politik 1958–1969', in *Kritische Studien zur Geschichtswissenschaft*, Bd. 149. Göttingen: Vandenhoeck & Ruprecht.

Prinz, L & A Pegels. 2018. 'The role of labour power in sustainability transitions: Insights from comparative political economy on Germany's electricity transition', *Energy Research & Social Science*, vol. 41, pp. 210–219.

Prognos AG. 2015. *Lehren aus dem Strukturwandel im Ruhrgebiet für die Regionalpolitik*. www.bmwi.de/Redaktion/DE/Publikationen/Studien/lehren-strukturwandel-ruhrgebiet-regionalpolitik.pdf?__blob=publicationFile&v=8 (last accessed June 2019).

RBB (Rundfunk Berlin-Brandenburg). 2018. 'Wirtschaftsminister: Braunkohle-Ausstieg nicht vor 2030', *rbb-exklusiv*, 23 June. www.rbb24.de/wirtschaft/beitrag/2018/06/wirtschaftsminister-peter-altmaier-braunkohle-ausstieg-lausitz.html (last accessed April 2019).

Reitzenstein, A. 2018. *A Climate for Ambition?* E3G – Third Generation Environmentalism.

—— & R Popp. 2019. *The German Coal Commission: A Role Model For Transformative Change?* E3G – Third Generation Environmentalism.

—— & S Schulz. 2018. *Deficit of Ambition: How Germany's New Coalition Government Falls Short on Climate & Energy*. E3G – Third Generation Environmentalism.

Renn, O & JP Marshall. 2016. 'Coal, nuclear and renewable energy policies in Germany: From the 1950s to the "Energiewende"', *Energy Policy*, vol. 99, pp. 224–232.

Schulz, S & J Schwartzkopff. 2016. *Instruments for a Managed Coal Phaseout: German and International Experiences with Structural Change*. E3G – Third Generation Environmentalism.

Schulze, S. 2018. 'Deutschland muss im Klimaschutz wieder eine Vorreiterrolle übernehmen. Interview by Jens Tartler & Nora Marie Zaremba', *Der Tagesspiegel*, 23 April. www.tagesspiegel.de/politik/umweltministerin-schulze-deutschland-muss-im-klimaschutz-wieder-eine-vorreiterrolle-uebernehmen/21200576.html23 (last accessed April 2019).

Statistik der Kohlenwirtschaft e.V. 2019. *Belegschaft im Steinkohlenbergbau, Gesamtbelegschaft nach Revieren ab 1945, Stand Januar 2019*. https://

kohlenstatistik.de/files/gesamtbelegschaft_reviere.xlsx (last accessed April 2019).

——. 2018. *Beschäftigte im Braunkohlenbergbau nach Revieren ab 1929, Stand Juli 2018.* https://kohlenstatistik.de/files/beschaeftigte.xls (last accessed April 2019).

Umweltbundesamt. 2018. 'Indicator: Greenhouse gas emissions'. www.umweltbundesamt.de/en/indicator-greenhouse-gas-emissions#textpart-1 (last accessed April 2019).

Verein der Kohlenimporteure. 2018. 'Jahresbericht 2018: Fakten und Trends 2017/18'. www.kohlenimporteure.de/publikationen/jahresbericht-2018.html (last accessed June 2019).

ZEIT ONLINE. 2018a. 'SPD-Chefin kritisiert Klimapolitik der Grünen', 31 August. www.zeit.de/politik/deutschland/2018-08/andrea-nahles-klimaschutzpolitik-die-gruenen-vorwurf (last accessed April 2019).

——. 2018b. 'Tausende Beschäftigte demonstrieren gegen Kohleausstieg', 24 October. www.zeit.de/wirtschaft/2018-10/braunkohle-demonstration-arbeitnehmer-kohlekommission-gewerkschaft-kohleausstieg (last accessed April 2019).

——. 2018c. 'Mehrheit der Deutschen gegen Rodung des Hambacher Forsts', 19 September. www.zeit.de/gesellschaft/2018-09/zivilgesellschaft-rodung-hambacher-forst-stopp-umfrage (last accessed April 2019).

8

A top-down transition: A critical account of Canada's government-led phase-out of the coal sector

Hadrian Mertins-Kirkwood & Ian Hussey

INTRODUCTION

As a fossil fuel superpower that positions itself internationally as an environmental leader, Canada occupies a conflicted place in the global fight against climate change. On the one hand, the country sees itself as an inspiration and a model to the world. Although historically obstructionist in international fora, a new government in 2015 made it a priority for Canadian officials to champion bold international action to transition to a clean economy. Canadian Environment Minister Catherine McKenna was instrumental in pushing for the Paris Agreement's aspirational 1.5°C target in 2015 and helped launch the Powering Past Coal Alliance in 2017, while Prime Minister Justin Trudeau emerged as the face of a new generation of climate-conscious political leaders. On the domestic front, the Canadian Government has implemented several milestone environmental policies in recent years, including a country-wide carbon pricing system. Subnational governments have done even more to reduce emissions and to transition toward a lower-carbon economy.

On the other hand, Canada remains an unabashed producer and profligate consumer of fossil fuels. On a per capita basis, Canadians are among the most energy-intensive and polluting people on the planet. Canada's climate targets are some of the weakest in the world and current policies to reduce emissions are inadequate to meet even that low bar. Those weak climate policies are directly undermined by government efforts to support the fossil fuel sector through a variety of direct and indirect subsidies. The same Justin Trudeau who says 'there is no hiding from climate change' maintains that 'no country would find 173 billion barrels of oil in the ground and leave them there' (The Canadian Press, 2016; *Maclean's*, 2017). In a symbolic rebuke of its climate credentials, the Canadian federal government spent $CAD 4.5 billion in 2018 to purchase a pipeline for transporting oil from the Alberta tar sands to the Port of Vancouver.

Nowhere are the contradictions and challenges of Canada's clean energy transition more apparent than in the province of Alberta – home to the oil sands and the majority of Canada's fossil fuel industry – where a major effort is underway to phase out the use of coal-fired electricity generation. Under provincial and federal regulations, all conventional power plants will stop burning coal by 31 December 2029, which will result directly in plant closures as well as indirect closures of coal mines across the country. From the perspective of reducing greenhouse gas emissions, the coal phase-out is a necessary and significant achievement consistent with Canada's climate policy goals. However, for the people who live and work in Alberta's many coal communities, the policy is costly and antagonising.

Recognising the political challenges – and responding to pressure from labour unions and social and environmental activists – the Alberta provincial government and the Canadian federal government are implementing policies to support coal workers and communities negatively affected by the phase-out. Employing the language of just transition advocates, these governments promise to make the transition off coal a fair one for workers in Canada and, by way of example, for workers around the world. Canada's centrally planned coal transition is a notable example of a just transition managed by the state, rather than through grass-roots activism, labour unions, or the private sector.

In this chapter, we explore the origins and design of the Canadian coal transition with an emphasis on the policies put in place to support coal workers and communities in Alberta. The first section outlines the economic role of the coal industry in Canada to determine the stakes of the transition. Among other issues, we discuss the number of jobs at risk from the closure of coal plants and coal mines. The second section investigates the Albertan case study in more detail by breaking down the transition policies put in place by the provincial government. We discuss the strengths and weaknesses of different policy approaches as well as their potential impacts. A concluding section summarises our main findings and discusses the broader implications of the Alberta coal transition. We consider the lessons and pitfalls of the Canadian example for other countries or subnational jurisdictions that may be considering or undergoing energy transitions of their own.

Overall, we conclude that Canada's policy approach for phasing out coal while supporting coal workers and communities is reasonable and pragmatic. Alberta's energy transition may be politically treacherous, but by supporting workers and communities it looks likely to succeed on most fronts. Nevertheless, we find that Canada's narrow focus on the coal industry distracts it from the need to phase out oil and gas production, which are ultimately more important for mitigating the country's climate impacts. While the Alberta case establishes a practical model for just transitions in other jurisdictions, this approach alone is insufficient for achieving a productive, inclusive and large-scale transformation of the energy system.

THE SOCIAL AND ECONOMIC CONTEXT OF CANADA'S COAL TRANSITION

Generally speaking, the rationale for phasing out coal-powered electricity generation is as follows:

1. Anthropogenic climate change is primarily caused by greenhouse gas (GHG) emissions from the combustion of fossil fuels;
2. The global economy must transition to lower-emission energy sources to avoid the worst effects of climate change;
3. The most emissions-intensive fuel is coal; therefore,
4. Efforts to fight climate change by reducing GHG emissions should start by replacing traditional coal power with other forms of energy.

Following that logic, the province of Ontario was the first Canadian jurisdiction to phase out coal power when it closed five generating stations between 2001 and 2014. Alberta followed suit when it announced in 2015 a phase-out of coal-fired electricity generation by 2030. At the time, five of the top ten industrial emitters in that province were coal plants (Thibault & Read, 2016). The federal government, which had announced a series of increasingly stringent emissions regulations for coal plants between 2011 and 2014, announced in 2016 that, with some small exceptions, a nationwide phase-out of coal power would be implemented by 31 December 2029, to reinforce and expand upon earlier provincial efforts. In each case, Canada's federal and provincial governments promised to fill the electricity generation gap with lower-emitting power sources and energy efficiency measures.

Factors other than climate change were at play for these governments, such as the health costs associated with air pollution. In Alberta, for example, the prevalence of coal power has resulted in some of the worst air quality among Canadian cities (Thibault & Read, 2016). The health costs alone of Alberta's poor air quality from burning coal for electricity have been estimated at $CAD 300 million per year (Anderson et al.,

2013). Nationally, poor air quality costs Canada in the order of $CAD 39 billion per year when all sources of pollution are included (Smith & McDougal, 2017). Nevertheless, without the climate imperative, Canada's regulatory coal phase-out would not have been implemented so comprehensively on such a short timeline (see, for example, Environment and Climate Change Canada, 2016).

Whether the recent federal and provincial coal phase-out policies are actually effective or significant from an emission reduction perspective is complicated. First, many of Canada's coal-fired electricity generators will be converted to natural gas-fired generators rather than shut down completely. Substituting one fossil fuel for a slightly cleaner one may reduce overall emissions, but not by nearly as a much as a wholesale transition to renewable energy sources. More importantly, Canada's coal sector plays a much smaller role than the oil and gas extraction industry both in terms of greenhouse gas emissions and as a contributor to the national economy. For Canada to reduce its emissions at a meaningful scale will require the country to reckon with oil and gas production, which has so far evaded serious regulatory attention by virtue of being cleaner than coal. We explore these caveats in more detail in the conclusion.

Nevertheless, the social and economic costs of the coal transition for those workers and communities dependent on the coal industry are clear. Canada's federal and provincial governments acknowledged as much when their phase-out policies were introduced. For example, in announcing its 31 December 2029 coal phase-out target, the federal government promised to 'work with provinces and labour organizations to ensure workers affected by the accelerated phase-out of traditional coal power are involved in a successful transition to the low-carbon economy of the future' (Environment and Climate Change Canada, 2016). Similarly, the Alberta Government promised to 'provide transition help to ... people working in the coal industry' as part of its broader climate policy plans (Government of Alberta, 2015a). In effect, Canada's federal and provincial governments promised a just transition of the coal sector, and they have since introduced tangible policies in pursuit of that goal.

Are they succeeding? To assess the adequacy of the transition policies put in place by Canadian governments, we must first establish the scope of the coal transition challenge. We begin by outlining the economic role of coal in Canada, including its contribution to economic growth and the labour market. We then outline the demands for a just transition from workers and communities that pushed governments to act on this issue.

Economic role of the Canadian coal industry

The Canadian coal industry includes consumers (coal-fired power plants) and producers (coal mines) of coal, both of which are affected by the coal phase-out. Although the federal and provincial regulatory phase-outs only apply to power plants directly, reduced demand will indirectly affect coal mines.

Coal is still used as a fuel source, either in whole or in part, at 15 power-generating facilities in Canada. A number of other coal plants have already shut down or been converted to alternative fuel sources across the country, such as the five plants in the province of Ontario closed or converted before 2014. The majority of remaining coal plants are clustered in Western Canada and in Atlantic Canada (see Table 8.1).

Table 8.1 Active coal-fired power plants in Canada (2018)

Facility name	Location	Net capacity from all fuel sources(megawatts)*
Sundance Power Plant	Wabamun, Alberta	1,861
Genesee Generating Station	Genesee, Alberta	1,376
Keephills Power Plant	Wabamun, Alberta	790
Sheerness Thermal Generating Station	Hanna, Alberta	780
Battle River Generating Station	Battle River, Alberta	689
Boundary Dam Power Station	Estevan, Saskatchewan	672
Lingan Generating Station	Lingan, Nova Scotia	620
Poplar River Power Station	Coronach, Saskatchewan	582
Belledune Thermal Generating Station	Belledune, New Brunswick	490
Brandon Coal Power Plant Canada	Brandon, Manitoba	327
Trenton Generating Station	Trenton, Nova Scotia	307
Shand Power Station	Estevan, Saskatchewan	276
Point Aconi Generating Station	Point Aconi, Nova Scotia	171
Point Tupper Generating Station	Point Tupper, Nova Scotia	154
HR Milner Generating Station	Grande Cache, Alberta	150

*   Capacity figures were collected by the authors in December 2018 from the official websites of Atco Power, Capital Power, Manitoba Hydro, Maxim Power, New Brunswick Power Corporation, Nova Scotia Power, SaskPower and TransAlta and corroborated by the national figures (Natural Resources Canada, 2018).

Coal-fired generating units provide 8,159 megawatts (MW) of generating capacity at these facilities with natural gas and other secondary sources bringing total capacity to 9,245 MW (Natural Resources Canada, 2018). In 2017, these plants consumed 33 million metric tonnes of coal to produce 56 million megawatt hours of electricity, which accounted for 9 per cent of total Canadian electricity generation (Statistics Canada, 2018a,b). However, in the coal-burning provinces of Alberta, Saskatchewan and Nova Scotia, coal provides between 45 and 50 per cent of total power.

The direct economic and employment impacts of coal-fired electricity generation are difficult to measure because power plants are typically owned by public or private utilities with multiple facilities using a mix of fuel sources. Electricity generation accounts for about 2 per cent of the Canadian economy overall, so coal power, given its minority share of total electricity generation, can be expected to account for a fraction of a per cent of gross domestic product (GDP), even in the coal-burning provinces (Statistics Canada, 2018c).

The employment figures are somewhat clearer for single-fuel facilities in small, relatively isolated communities such as Estevan, Saskatchewan, where the Shand and Boundary Dam power stations together account for about 400 jobs, or Cape Breton, Nova Scotia, where the Lingan, Point Aconi and Point Tupper stations together account for about 300 jobs.[1] Extrapolating from these cases, a reasonable estimate for total coal power plant jobs across Canada is in the range of 3,000–5,000 people. The majority of those jobs are in Alberta, especially in the region surrounding the city of Edmonton where the massive Sundance and Keephills plants are located.

On the producer side, there are 19 coal mines operating in Canada with the majority located in the western provinces of British Columbia and Alberta (Coal Association of Canada, n.d.). Measuring the economic impact of these mines is much simpler than power plants because coal mining is a distinct category in the statistical accounts. In total, those mines accounted for $CAD 3.5 billion in economic activity in 2017, of which $CAD 539 million came from Alberta (Statistics Canada, 2018c,d). In the same year, the coal-mining industry employed 7,700 people of which 2,325 worked in Alberta (Statistics Canada, 2018e). As a share of the overall economy, those figures are relatively small. Coal mining accounts for less than 0.2 per cent of total GDP and less than 0.1 per cent of total employment nationally.

The regulatory phase-out of coal-fired electricity generation will negatively affect coal power plants, which will in turn reduce demand from coal mines. However, for a number of reasons, the policy will not neces-

---

1   Authors' calculations using North American Industry Classification System (NAICS) code 2211 (Electric power generation, transmission and distribution) in the census agglomerations of Estevan and Cape Breton (Statistics Canada, 2017).

sarily result in the closure of all coal mines and power plants or the loss of all coal-related employment.

First, the coal phase-out policy does not require plants to close completely as long as they transition from coal to other fuels. When the province of Ontario phased out coal power between 2001 and 2014, the provincial utility converted some of its existing coal-firing facilities to use biomass instead (Ontario Power Generation, n.d.). Looking ahead, as we discuss in the next section, many existing power plants will shift from coal to natural gas (see, for example, TransAlta, n.d.). Gas-fired plants only require a third of the labour of coal-fired plants so job losses are to be expected, but facilities will remain open and a portion of workers will still be employed in those communities (Vriens, 2018:13).

Second, only thermal coal, which is used for electricity generation, is affected by the coal phase-out policy. Metallurgical coal, which is mainly used in steel production, will mostly be unaffected. Nearly half the coal produced in Canada is metallurgical, so while many mining jobs may be lost, not all coal workers are at risk from present policies (Statistics Canada, 2018f).

It is also worth noting that even where coal facilities do shut down in the coming years, the coal phase-out policy is unlikely to be the only cause. First, the Canadian thermal coal industry was in decline even before the latest phase-out regulations were introduced. Since peaking in the early 2000s, economic output from coal mining has declined 40 per cent nationally and more than 70 per cent in Alberta (Statistics Canada, 2018c). The coal industry has historically been subject to a boom-and-bust cycle in response to global commodity prices. Over the past decade, coal prices have fallen significantly as many countries shift to cleaner alternatives, which has negatively impacted the Canadian industry.

Second, the rapidly falling price of renewable energy is starting to displace coal power even without regulatory intervention (Landberg & Hirtenstein, 2018). As coal plants reach the end of their useful life, the economic incentive to refurbish or expand them is diminishing. In sum, the coal industry, which includes both power plants and mines, accounts for a relatively minor share of Canada's economy (a fraction of a per cent of GDP) and a correspondingly small number of direct jobs (in the range of 10,000–13,000 workers or 0.1 per cent of the national workforce). In comparison, the oil and gas extraction industry – Canada's largest single sector in terms of greenhouse gas emissions – accounts for 5 per cent of GDP and upwards of 1 per cent of employment (Statistics Canada, 2018c,e). Moreover, the 'phase-out' of coal does not necessarily mean the end of either coal mines or coal power plants, since many are engaged in the production of metallurgical coal and the consumption of alternative fuel sources, respectively. Potential job losses in the coal sector are thus mitigated. In Alberta, for example, approximately 2,000 jobs are considered at risk from the phase-out, even though at least 3,000 people work in the current coal sector (Vriens, 2018:13; AFL & CTC, 2017:6).

Community and labour demands for a just transition

The phase-out of the coal industry may have a limited economic impact for Canada overall, but the stakes are much higher in those communities where facilities are at risk. Most coal power plants and nearly all coal mines are in rural regions where those facilities are responsible for a substantial portion of employment and investment. Coronach, Saskatchewan, for example, is highly dependent on the Poplar River station and its potential closure is causing 'frustration, fear, and worry among local residents' (Fisher et al., 2018:14). In Alberta, there are approximately twenty municipalities and First Nations impacted by the coal phase-out. Most of them are rural communities, although a few larger coal towns are bedroom communities of the provincial capital, Edmonton.

Opposition to the coal phase-out has manifested in two distinct ways. On the one hand, the policy has entrenched resistance to climate policies and stoked support for populist politicians, mostly from conservative parties (see, for example, Graney, 2018). On the other hand, for those workers and communities that accept the environmental necessity of ending coal power, the policy has spurred demands for more active government support through the transition.

Demands for a just transition to a cleaner economy in Canada pre-date the coal phase-out. The Canadian Labour Congress, Unifor, and other major labour organisations have, for decades, demanded transitional supports for fossil fuel workers in response to climate measures (Mertins-Kirkwood, 2017:4). Social justice organisations, environmental groups and non-union worker groups have also advocated for a government-supported transition from fossil fuels to renewable energy sources (see, for example, Cooling et al., 2015; MacArthur et al., 2016). However, it was the concrete introduction of policies to phase out coal that provoked a more focused response from just transition advocates in Canada.

The Alberta Federation of Labour (AFL) has played a pivotal role in responding to and shaping the coal transition conversation in that province. Shortly after the coal phase-out was announced, the AFL convened a coalition of labour unions representing workers in the coal industry to develop a just transition framework for the province (AFL & CTC, 2017). In March 2017, the coalition proposed a series of measures to promote a productive transition to a cleaner economy, including programmes to retrain and counsel workers for new jobs, programmes to facilitate job transfers and relocation, and funds for economic diversification. The province responded by introducing a set of transition programmes for workers and communities in the autumn of 2017, which we discuss in the next section.

At the national level, pressure from labour unions spurred the federal government in December 2017 to announce a Task Force on Just Transition for Canadian Coal Power Workers and Communities, which submitted its recommendations to the government in late 2018 (Environment and Climate Change Canada, 2018a). The task force recommended increasing financial supports to affected workers, investing in economic diversification in affected areas, and protecting those policies through legislation. The government responded with the modest first step of opening two transition centres in Albertan coal communities, which are intended to act as community hubs for laid-off workers by providing career counselling and other services (Western Economic Diversification Canada, 2018). Further initiatives are forthcoming.

Canadian environmental organisations and social justice organisations have actively supported labour unions' demands for a just transition. Unlike in the United States and some other countries grappling with similar issues, there are few tensions between the Canadian environmental and labour movements on the issue of just transition. Climate Action Network Canada, for example, has mobilised its dozens of member organisations to pressure governments in support of workers (Climate Action Network Canada, 2018). However, these groups play a clearly subordinate role on this issue by allowing labour unions to be the voice and face of the just transition movement.

In sum, the potential economic costs of the Canadian coal phase-out spurred labour organisations to demand just transition programmes for coal workers and communities from the federal and provincial governments. In response, those governments have acknowledged that phasing out coal has direct costs for certain people and regions of the country and, consequently, that governments have a responsibility to support affected communities during the transition. In the next section, we evaluate the adequacy of Alberta's transition policies given the scope of the challenge described above.

CASE STUDY: THE ALBERTA COAL TRANSITION

In 1995, then-Premier Ralph Klein began to deregulate Alberta's electricity market. By 2001, the province's energy market was fully deregulated and all of the power plants in the province were privately owned (Wallace, 2001). The private energy market has had a number of important consequences, including excess supply. In 2016, Alberta's power-generating capacity was more than 35 per cent above the supply needed to meet peak demand (Alberta Electric System Operator, 2018). Importantly, very little of this new capacity was from renewable energy sources. Although the prevalence of coal-fired generation in Alberta's electricity market declined from over 80 per cent of installed capacity in the late 1980s to approximately half in 2015, natural gas-fired generation in the province increased from about 10 per cent of total production in 1988 to almost 40 per cent in 2015 (AFL & CTC, 2017:9).

In May 2015, the left-of-centre Alberta New Democratic Party (NDP) led by Rachel Notley won a majority government, ending the 44-year political dynasty of the right-wing Progressive Conservative (PC) party. Acknowledging the province's continued dependence on fossil fuels for electricity – and breaking from the province's historical foot-dragging on climate action and environmental protection – the NDP platform included commitments to phase out coal-fired electricity generation and to expand wind and solar energy production. The new government promised to reduce the province's dependence on fossil fuels in general and to move away from fossil fuel-based electricity generation in particular. Politically, the excess power supply made the phase-out of coal units easier to start than if the province had a tighter reserve margin and thus higher electricity prices (Vriens, 2018:6).

Alberta's Climate Leadership Plan, announced six months into the NDP's mandate, included a carbon levy on transportation and heating fuels; an annual oil sands emissions cap of 100 megatonnes of carbon dioxide equivalent (CO_2e); a directive to increase renewable energy in the province from 9 per cent of total power generation in 2015 to 30 per cent by 2030; the creation of Energy Efficiency Alberta, a publicly owned corporation that promotes and channels public funds toward energy efficiency; and the phase-out of coal-fired electricity generation by 2030 (Government of Alberta, 2015b).

In the three years that followed, the Alberta NDP and federal Liberal governments implemented several policies and related financial commitments to enable the coal phase-out to succeed. The two governments have also worked hard to 'own' the coal phase-out politically. However, outside of Ontario, the Canadian coal phase-out truly began in September

2012 with the announcement of new regulations for coal-fired electricity units by then-Prime Minister Stephen Harper's Conservative Government (Environment and Climate Change Canada, 2012). For Alberta, the province that has produced the most coal-fired electricity in recent years, the 2012 federal regulations meant twelve of the province's 18 coal units (located at three of the province's six coal plants) would close by the end of 2030 anyway. The Harper Government did not develop a transition plan for affected workers, communities, or businesses, and neither did the then-PC Government of Alberta.

The Notley Government's climate plan only affects the province's six other, younger coal units (at three separate coal plants), the youngest of which could have operated until 2061 under the 2012 Harper regulations. Three publicly traded corporations – TransAlta, ATCO and Capital Power – own the six units, and they will convert them to natural gas-fired units in the next decade. In other words, the 2015 federal phase-out regulations made little practical difference in Alberta because every coal-fired generating unit was already slated for phase-out under pre-existing federal and provincial policies.

In addition to the direct regulatory phase-out of coal-fired units, the Alberta Government announced two other major changes to the provincial energy market. In November 2016, the government announced its intentions to transition the province from a lightly regulated, private energy market to a new model that includes both an energy market and a capacity market (Government of Alberta, 2016a, 2018a). A capacity market requires producers to compete to sell power and to compete for payments to provide electricity capacity on demand, both of which will encourage new capital investments by electricity firms. The system will come into effect by 2021.

In June 2017, the NDP government also implemented an electricity price cap of 6.8 cents per kilowatt hour (Government of Alberta, 2017a). The price cap, which will last until 31 May 2021, is intended to protect families and businesses from possible price volatility as the province transitions to the new capacity market. The 2002–18 average price for electricity in Alberta was 7.3 cents, with a historic high of 15.06 cents in January 2012 and a historic low of 2.88 cents in April 2017, so the new policy provides a significant degree of short-term price stability.

Taken together, Alberta's policies are designed to facilitate a smooth transition of the electricity sector from a coal-burning, private market to a more heavily regulated, cleaner electricity grid. The costs, as discussed in the previous section, will be felt by coal companies, coal workers and coal communities. The province has responded by implementing transition programmes aimed at those three groups. We discuss each in turn.

Transitioning Alberta's coal-fired power corporations

As a matter of law, the province did not have to compensate coal companies because of the introduction of new regulations (Wilt, 2017). Moreover, four of the six coal-fired units affected by the Alberta Government's accelerated coal phase-out would receive a fair return on capital by 2030, thus calling into question the case for compensation from the provincial government (Marr-Laing & Thibault, 2015). The other two generating units, Keephills 3 and Genesee 3, are both co-owned by TransAlta and Capital Power, and these companies were aware when they commissioned these coal units in 2005 and 2011 respectively that their investments would be negatively affected by climate policy in their lifetimes.

Nevertheless, from the outset, Premier Notley instructed the Minister of Energy to avoid stranding any capital in the coal phase-out (Government of Alberta, 2015a). The government was concerned about maintaining investor confidence because the province requires approximately $CAD 15 billion in private investments to replace the lost coal capacity with (converted) gas-fired units and a further $CAD 10.5 billion in private investments to increase renewable generation to 30 per cent of total electricity production by 2030. TransAlta, ATCO and Capital Power are among the province's largest power-generating corporations, and they will likely continue to own a sizeable portion of Alberta's installed capacity as the province moves to a 70/30 mix of gas-fired and renewable generation in the next decade.

In November 2016, the province announced a set of Off-Coal Agreements (Government of Alberta, 2016b). The agreements will see the Government of Alberta compensate TransAlta, ATCO and Capital Power a total of $CAD 1.36 billion over 14 years. The funds come from the province's carbon tax for large industrial emitters, which was first implemented by the former PC Government in 2007. The compensation was based on the six coal units' '2015 net book value, pro-rated by the number of years between 2030 and the original federal end-of-life date (the years that the unit was "stranded" by Alberta's policy)' (Vriens, 2018:11).

The Off-Coal Agreements are confidential, but some general details have been made public. For example, the three companies agreed to keep their headquarters and a nominal number of employees in Alberta, and to keep investing in the province's power system. Six months after the Off-Coal Agreements were signed, in May 2017, ATCO and TransAlta, the two biggest coal-fired electricity producers in Alberta, announced they would convert their coal units to natural gas well before the December 2029 deadline. ATCO plans to convert its units by 2020 and

TransAlta will do the same by 2022 (Morgan, 2017). Capital Power, on the other hand, has said it will continue to operate its coal units until the government-mandated deadline, when the units will be converted to gas (Kent, 2016).

Because the provincial and federal governments adopted clear rules and timelines, companies are able to plan ahead to figure out the optimal timing to phase out their coal units, invest in renewables, and build other new infrastructure, like gas pipelines to their converted gas-powered units. TransAlta, for example, says it will save $CAD 1.5 billion by converting its coal units to gas by 2022. The company is taking advantage of the fact that Alberta's electricity market is significantly oversupplied, natural gas is an inexpensive input, and the carbon cost for gas is lower than that of coal. Similarly, ATCO's conversion to gas power by 2020 is motivated by the expiry of their power-purchase contracts in 2020 and by the availability of cheap natural gas, as opposed to the province's industrial carbon tax that they have been paying since 2007.

At least 13 of Alberta's 18 coal units will be converted to gas by 2029 while one additional conversion is likely to be announced. Coal-to-gas conversions are more economical for Alberta than the construction of a whole new fleet of natural gas combined-cycle plants (Boston, 2016). Companies investing in new gas plants would expect them to operate for at least thirty years, which could make the firms and the province vulnerable to future federal regulations on gas-fired plants, cheaper renewables, or international pressure. According to the new federal regulations, coal-to-gas units can only operate 5–10 years past their federal end-of-coal life, depending on their GHG emission profile (Government of Canada, 2018). Alberta's coal-to-gas conversion will reduce overall GHG emissions from the electricity sector but, for Canada to meet its emissions targets, all the new gas-fired capacity will need to be replaced by renewables in the 2030s.

Workforce Adjustment Committees that involve companies, labour unions, government representatives and other stakeholders operate and will continue to operate at worksites undergoing the transition from coal to natural gas to support the transition of workers to another job, retraining, or retirement as needed. Collaborations between the private sector and organised labour contribute to a just transition in the province, but these programmes are much more limited in scope and scale than the transition policies put in place by governments.

Overall, the preferential treatment of energy corporations in Alberta's coal transition reflects the political economy of the province's energy market. The province is almost wholly dependent on a handful of private firms for an essential service, which ensured they had significant leverage in their negotiations with the provincial government (Vriens, 2018:9). Not only did the power companies secure generous compensation and lenient terms for the coal phase-out, but they were also guaranteed the right to convert most coal-fired generators to gas-fired generators. Coal-to-gas conversion is more economical (and profitable) in the short term because it requires less new infrastructure than installing the equivalent capacity in renewables, but pursuing this option requires the province to make a continued commitment to fossil fuels even as climate science demands a more radical transition to clean energy. Canadian environmentalists have long made the argument that natural gas is neither viable or necessary as a transition fuel to renewables (Bramley, 2011:36). If Alberta's power plants were owned by a public utility, as they are in most other provinces, the government would have had more latitude to pursue a complete fossil fuel phase-out of the electricity sector.

Alberta's transition programmes for coal workers

The coal phase-out in Alberta will likely see more than 2,000 workers laid off, and these lay-offs will probably be concentrated in two periods – rolling lay-offs for 2018–22 and for 2028–29 – because of the three electricity firms' coal-to-gas conversion plans.

In September 2016, the Alberta Government established an advisory panel to consult with coal workers and communities – including municipal and First Nations leaders, small businesses, and community economic development organisations – to examine potential impacts of the coal phase-out and to identify ways to support worker transition (Government of Alberta, 2016c). The advisory panel's recommendations to the government were released the following year (Government of Alberta, 2017b). Shortly thereafter, the Notley Government announced the creation of a $CAD 40 million fund to support several transition programmes for the province's coal workers (Government of Alberta, 2017c,d).

During the advisory panel's consultation period, the Alberta Federation of Labour-led Coal Transition Coalition established itself as an effective advocate for both coal workers and their communities. The coalition hosted several town hall meetings in coal communities and published a report in March 2017 that includes four workforce transition case studies to identify best practices and lessons that might be useful for Alberta's transition away from coal power (AFL & CTC, 2017). Many of the recommendations from the coalition's research were mirrored in the recommendations submitted to the provincial government by its advisory panel.

The Notley Government's financial, employment and retraining programmes for workers began to operate in January 2018 (Government of Alberta, 2017d). The transition programmes consist of the following six components:

1. The bridge to re-employment relief grant provides workers 'up to 75% of their previous weekly earnings when combined with [federal] Employment Insurance benefits' (ibid.). This relief grant can last up to 45 weeks or until the worker starts new full-time employment.
2. The bridge to retirement relief grant provides financial support for workers that are close to retirement but not yet eligible for their employer pension. This grant provides 'up to 75% of their previous weekly earnings for up to 72 weeks, or receipt of pension, or when gross employment income is greater than the relief payment, whichever is shorter' (ibid.).
3. Coal workers that are laid off and move at least 40 kilometres to begin a new job are eligible to be reimbursed up to $CAD 5,000 for moving-related expenses.
4. Laid-off coal workers are eligible for a tuition voucher of up to $CAD 12,000 if they wish to return to school to retrain within five years of being laid off.
5. Career consultants and employment service providers are made available to work directly with coal workers to share information, to develop individualised plans, and to provide short-term skills development courses as needed.
6. The provincial government provided employers and unions with a list of qualified facilitators who can be hired to assist employers, workers and unions with setting up a workforce adjustment committee to create a tailored transition plan for individual worksites based on labour market research.

None of the programmes includes a job creation component, although economic diversification is an element of the community transition programmes discussed in the next section.

Initial uptake of the coal worker programmes has been modest. Official figures on the number of coal workers applying for benefits are not yet publicly available, but the Alberta Ministry of Labour provided the following comment in response to researchers' questions (Hussey & Jackson, 2019):

Through August 30, 2018 there have been 68 approved Bridge to Re-employment grants and 24 Coal and Electricity Transition Tuition Vouchers. As of October 1, 2018 only one of the affected coal companies has announced layoffs. With the Alberta economy improving, that company has seen significant attrition. This program is new and the uptake seems to align with the number of layoffs that have been reported. (Personal communication, 30 October 2018)

In total, during the first eight months of their existence, the Government of Alberta's programmes supported 92 coal workers bridging to a new job or returning to school. With the provincial economy improving, it is not surprising that electricity firms are losing employees to attrition (that is, workers are leaving for better employment opportunities of their own accord). Many of the workers who would be eligible for transition supports have skills that are in demand in other sectors of the economy.

In other words, many coal workers are transitioning into new industries as the economy grows independent of the coal phase-out. This trend is unlikely to stop, especially as alternative electricity industries scale up. Indeed, Alberta is forecast to see a net gain of 29,000–38,000 jobs from 2017 through 2029, or about 2,200–2,900 full-time equivalent jobs per year, because of the transition away from coal and toward gas-fired and renewable generation. Job gains are projected from the following three sources:

1. To triple renewable energy production by 2030, the Government of Alberta is working to attract $CAD 10.5 billion in private investment, which is expected to create 7,200 job-years over 13 years or 550 full-time equivalent jobs per year (Government of Alberta, 2017e).

2. The coal-to-gas conversions are expected to require $CAD 15 billion in private investment (Vriens, 2018:9), and this investment could create up to 150,000 person-years of employment or 15,000 full-time equivalent jobs over the course of a decade (AFL & CTC, 2017:22).

3. The AFL-led Coal Transition Coalition estimates that Alberta should receive roughly $CAD 2 billion of the $CAD 21.9 billion promised by the federal government to help get Canada's electricity system to 90 per cent non-emitting by 2030. The coalition asserts that '$2 billion could attract capital investments on the order of $10 billion to $20 billion, or on the order of another 9,000 to 18,000 jobs over that 11-year period' (AFL & CTC, 2017:22).

The shift to new energy sources is not all good news for workers, since the power plant and mine jobs being lost are mostly unionised and the new jobs are mostly short-term construction positions, many of which will not be unionised. The potential erosion of unionisation rates in the energy sector is especially troubling given Alberta's historically low levels of unionisation. In 2014, before the NDP rose to power, Alberta's unionisation rate was 22.1 per cent, which was the lowest in the country and 8 percentage points lower than the national average (Statistics Canada, 2018g). Since the NDP won the 2015 Alberta election, the province has seen a bump in unionisation because of reforms made by the NDP to the provincial Labour Code and because the government refrained from slashing public sector spending during the 2015–16 oil-price-induced recession. Alberta's 2018 unionisation rate was 24.5 per cent, but in the private sector, which in Alberta includes power plants, the unionisation rate is closer to 10 per cent.

Nevertheless, while job quality and security remain a concern, Alberta's electricity transition alone is unlikely to drive net employment losses, even in many of the communities directly impacted by the coal phase-out.

Alberta's transition programmes for coal communities

Most of the near-term impacts of the phase-out on Alberta's thermal coal communities are the result of the federal government's 2012 regulations, not the Alberta NDP's 2015 climate plan. Yet the absence of a community transition plan from either the Harper Government in 2012 or the then-PC Government of Alberta meant affected communities had little support during the transition off coal. In late 2017, based on the recommendations of their advisory panel, the Alberta NDP Government created a $CAD 5 million Coal Community Transition Fund (Government of Alberta, 2017f). About twenty Alberta municipalities and First Nations impacted by the coal phase-out were eligible to apply to the fund.

In March 2018, the Notley Government announced that twelve projects in 17 coal communities had been funded, exhausting the $CAD 5 million fund. The projects include strategic planning, feasibility studies, tourism development, and 'work to expand economic hubs, including agribusiness, transportation and high-tech industries' (Government of Alberta, 2018b). Unfortunately, further details on the funded projects have not been made available by the government.

In addition to the Coal Community Transition Fund, the Alberta NDP Government created a second $CAD 30-million fund in 2017 over two years through the Community and Regional Economic Support program (Government of Alberta, 2017g). This fund was available to rural communities across the province, including coal communities.

More recently, in November 2018, Notley's Government announced an investment of $CAD 200 million over the next twenty years in response to growing demand for renewable energy (Government of Alberta, 2018c). The Community Generation Program will enable local communities, neighbours and institutions to participate in and benefit from renewable energy projects such as wind, solar, hydro and biomass.

Taken together, these three funds and the economic diversification and community electricity generation projects that they enable are helping to put a number of local communities in the province, many of them in rural areas, on a path toward economic and environmental sustainability. Some municipalities' budgets may see a decline in revenues from property taxes as a result of mine closures, electricity unit closures, or electricity unit conversions to natural gas, but the extent is not yet known. Many of the coal unit conversions and closures and the coal-mine closures will occur in the next four years, so further research will be needed in the coming years to evaluate the consequences.

IMPLICATIONS FOR FUTURE ENERGY TRANSITIONS

Canada's multi-pronged phase-out of coal-fired electricity generation will see the end of the thermal coal industry by 31 December 2029. The social and economic costs for the country on the whole are limited given coal's small share of electricity generation, but the regions currently reliant on coal for their power and livelihoods will be hit harder than others. In response, governments at the provincial and federal level have promised a just transition for coal workers and communities.

The province of Alberta, where the largest share of coal is burned, is leading the way with a series of programmes designed to ease the transition to a cleaner electricity system. Importantly, the provincial government consulted with labour unions and other community representatives to design a policy package that meets their needs. This collaborative approach between the state and organised labour is emblematic of the social reformist approach that has historically dominated the just transition discourse, rather than a more radical or grass-roots approach that might have seen community-led transition initiatives at its centre.

Whether Canada's state-led just transition programmes are ultimately effective in supporting coal workers and communities will become clearer over the coming decade, but by all indications they are on track to do so. In Alberta, the transition away from coal and toward gas-fired and renewable electricity production is projected to have a net employment benefit of approximately 2,200–2,900 full-time equivalent jobs per year. The favourable employment forecast suggests the provincial and federal governments have developed plans with affected workers, communities and businesses that will not only reduce greenhouse gas emissions and air contaminants but also create a substantial net economic gain. Importantly, while the coal phase-out is not being celebrated in affected communities, the state's just transition approach has mitigated resentment and moved the political conversation forward. Many affected workers and communities now frame the issue as one of how best to adapt to a changing energy industry, rather than debating the validity of the coal phase-out policy (Varcoe, 2017).

For other economies grappling with the social and economic costs of the energy transition, the Canadian case may serve as a useful model for mitigating harm from (and ensuring political buy-in for) environmental policies. The Alberta policy package is an especially good example of how workers and communities can help shape climate policies that affect them. Unfortunately, the value of the Canadian model for other energy transitions is complicated by three major caveats.

First, the scope of Canada's energy transition is extremely narrow from an economic perspective. Although coal's role is not insignificant, it accounts for a minority of national electricity generation and a vanishingly small share of national employment. Phasing out coal in Canada is simply not a transformational decision that requires ambitious transition policies. In contrast, many countries around the world – from Australia to South Africa to China to Poland – are far more dependent on coal (International Energy Agency Statistics, 2014). Places where coal (or any other fossil fuel) is a primary electricity source and major economic driver will require different approaches to transition (see Goods, Van Niekerk, Reitzenstein et al., and Snell, in this volume).

Second, both the phase-out of coal-fired electricity generation in Canada and the corresponding transition of coal workers into new industries is occurring largely independent of the government policies specifically targeted at the coal industry. In Alberta, for example, coal's cost competitiveness has been declining for over a decade because of low natural gas prices. The introduction of an industrial carbon tax in 2007 and a consumer carbon tax on transportation and heating fuels in 2017 further eroded the viability of coal power. Indeed, two of the three companies affected by that province's accelerated coal phase-out are speeding up the transition by almost a decade because it makes financial sense for them to do so, not because the government requires it.

On the employment side, many coal workers are changing jobs pro-actively and voluntarily, rather than subjecting themselves to any planned transition programme. As the Canadian economy grows and adapts, particularly in the alternative energy sector, jobs are being created that are absorbing skilled workers from coal mines and power plants. For other jurisdictions studying the Canadian model, the effectiveness of the regulatory coal phase-out and its accompanying just transition policies are difficult to separate from the underlying political and economic conditions driving changes in the electricity sector.

Third, and most importantly, Canada's coal phase-out is unlikely to have a significant impact on the country's overall GHG emissions picture. In part, the limited environmental benefits of the phase-out are due to the energy source being phased in to replace coal. Instead of non-emitting, renewable sources, the coal phase-out will see a large increase in the use of natural gas-fired generation. While natural gas produces fewer GHG emissions per unit of energy than either oil or coal, the production of natural gas releases methane, which is far more destructive than carbon dioxide in the short term. Regardless, all fossil fuels must be entirely phased out by mid-century to meet national and global emissions targets, so new investments into natural gas are misguided and problematic from the perspective of global climate change mitigation (Environmental Defence, 2018:4). Canada's embrace of coal-to-gas conversion sets a problematic precedent for coal-burning countries of comparatively lesser means that are being pressured to reduce their coal use.

At a more fundamental and structural level, however, the coal phase-out is unlikely to significantly reduce Canada's carbon footprint because coal already plays only a small role in the country's emissions profile. In 2016, the latest year for which data are available, Canada produced 704 megatonnes (Mt) of carbon dioxide equivalent, of which coal accounted for less than 10 per cent (Environment and Climate Change Canada, 2018b). In contrast, a quarter of emissions came from oil and gas production and another quarter came from the transportation of people and goods. The oil industry in particular has seen major growth in recent years with a corresponding environmental impact. Emissions from the production of oil and gas were 16 per cent higher in 2016 than in 2000. Over that period, Canadian production of crude oil increased 80 per cent, including a greater than 400 per cent increase in bitumen production from the oil sands.[2] Natural gas production and associated emissions have actually declined over the same period, in part due to competition from the US shale gas boom, but not nearly far enough to offset growth in the oil sector (Statistics Canada, 2018j).

In this context, the political focus on phasing out coal-fired electricity generation (and even the concerns about phasing in natural gas) is a distraction from the growth of Canada's outsized and highly polluting oil industry. Meeting Canada's emissions targets will require governments to tackle oil like they tackled coal, but the social and economic costs will be much greater, which explains in part their hesitation to do so.

Consequently, jurisdictions turning to Canada's coal phase-out as a model for just transition must recognise the limited ambition of the policy. Modest social programmes targeted at displaced workers may be effective when the energy transition is in its infancy and the social and economic stakes are relatively low. A transition that brings about transformative environmental justice, on the other hand, must do more to dismantle structural inequalities and put the interests of workers and communities above those of corporations. Such a transition will not be achieved without dismantling corporate power. So while representing a significant step in the right direction, Canada's coal phase-out still falls short when measured against the type of bold transformative action that the scale of the climate crisis demands.

REFERENCES

Alberta Electric System Operator. 2018. *Annual Market Statistics*. www.aeso.ca/download/listedfiles/2017-Annual-Market-Stats.pdf (last accessed April 2019).

AFL (Alberta Federation of Labour) & CTC (Coal Transition Coalition). 2017. *Getting it Right: A Just Transition Strategy for Alberta's Coal Workers*. Alberta Federation of Labour. www.coaltransition.ca/read_the_report (last accessed April 2019).

Anderson, K, T Weis, B Thibault, F Khan, B Nanni & N Farber. 2013. *A Costly Diagnosis: Subsidizing Coal Power with Albertans' Health*. The Pembina Foundation, The Asthma Society of Canada, The Canadian Association of Physicians for the Environment, The Lung Association, Alberta & Northwest Territories, and The Pembina Institute. https://ab.lung.ca/sitewyze/files/costly-diagnosis.pdf (last accessed April 2019).

Boston, T. 2016. *Letter to Premier Notley*. Government of Alberta. www.alberta.ca/documents/Electricity-Terry-Boston-Letter-to-Premier.pdf (last accessed April 2019).

2 Authors' calculation for the period 2000–15 (Statistics Canada, 2018h; Statistics Canada, 2018i).

Bramley, M. 2011. *Is Natural Gas a Climate Change Solution for Canada?* David Suzuki Foundation, The Pembina Institute & the Pembina Foundation. https://davidsuzuki.org/science-learning-centre-article/natural-gas-climate-change-solution-canada (last accessed April 2019).

Climate Action Network Canada. 2018. *Climate Action Network Canada Reacts to the Canadian Government's Announcement of the Just Transition Task Force*, 16 February. https://climateactionnetwork.ca/2018/02/16/climate-action-network-canada-reacts-to-the-canadian-governments-announcement-of-the-just-transition-task-force (last accessed April 2019).

Coal Association of Canada. (n.d.). 'Mine map. Careers in Coal'. https://careersincoal.ca/mine-map (last accessed April 2019).

Cooling, K, M Lee, S Daub & J Singer. 2015. *Just Transition: Creating a Green Social Contract for BC's Resource Workers*. Canadian Centre for Policy Alternatives British Columbia Office. www.policyalternatives.ca/publications/reports/just-transition (last accessed April 2019).

Environment and Climate Change Canada. 2018a. *Task Force on Just Transition for Canadian Coal Power Workers and Communities Terms of Reference*. Government of Canada. www.canada.ca/content/dam/eccc/documents/pdf/climate-change/canadian-coal-power-workers/canadian-coal-power-workers_en_03.pdf (last accessed April 2019).

——. 2018b. *National and Provincial/Territorial Greenhouse Gas Emission Tables*. Government of Canada. https://open.canada.ca/data/en/dataset/779c7bcf-4982-47eb-af1b-a33618a05e5b (last accessed April 2019).

——. 2016. *The Government of Canada Accelerates Investments in Clean Electricity*, 21 November. Government of Canada. www.canada.ca/en/environment-climate-change/news/2016/11/government-canada-accelerates-investments-clean-electricity.html (last accessed April 2019).

——. 2012. *Harper Government Moves Forward on Tough Rules for Coal-Fired Electricity Sector*, 5 September. Government of Canada. www.ec.gc.ca/default.asp?lang=En&n=714D9AAE-1&news=4D34AE9B-1768-415D-A546-8CCF09010A23 (last accessed April 2019).

Environmental Defence. 2018. *Canada's Oil & Gas Challenge: A Summary Analysis of Rising Oil and Gas Industry Emissions in Canada and Progress Towards Meeting Climate Targets*. Environmental Defence & Stand.earth. https://environmentaldefence.ca/report/canadas-oil-and-gas-challenge (last accessed April 2019).

Fisher, J, R Malena-Chan, & H Carlson. 2018. *Bridging the Gap: Building Bridges Between Urban Environmental Groups and Coal-producing Communities in Saskatchewan*. Climate Justice Saskatoon.

Government of Alberta. 2018a. *Electricity Capacity Market*. www.alberta.ca/electricity-capacity-market.aspx (last accessed April 2019).

——. 2018b. 'Helping coal communities diversify', 15 March. www.alberta.ca/release.cfm?xID=54590B0E0FC9A-A0C4-7860-C3B79E5673061864 (last accessed April 2019).

——. 2018c. 'Putting power in the hands of communities', 22 November. www.alberta.ca/release.cfm?xID=620855BF8D8B3-9B90-88C9-FD9330F8A9406D44 (last accessed April 2019).

——. 2017a. *Electricity Price Protection.* www.alberta.ca/electricity-price-protection.aspx (last accessed April 2019).

——. 2017b. *Supporting Workers and Communities: Recommendations to the Government of Alberta.* www.alberta.ca/assets/documents/advisory-panel-coal-communities-recommendations.pdf (last accessed April 2019).

——. 2017c. 'New transition supports for Alberta coal workers', 10 November. www.alberta.ca/release.cfm?xID=48946866B1DC2-B873-AEB5-FC07E467C571A5AF (last accessed April 2019).

——. 2017d. *Support for Workers Affected by Coal Phase Out.* https://www.alberta.ca/support-for-coal-workers.aspx (last accessed April 2019).

——. 2017e. 'New jobs, investments to come from renewables', 24 March. www.alberta.ca/release.cfm?xID=4653114186F9D-D960-29D3-F989FD2E0A0F0DF6 (last accessed April 2019).

——. 2017f. *Coal Community Transition Fund.* www.alberta.ca/coal-community-transition-fund.aspx (last accessed April 2019).

——. 2017g. *Community and Regional Economic Support (CARES) Program.* www.alberta.ca/community-regional-economic-support-program.aspx (last accessed April 2019).

——. 2016a. 'Consumers to benefit from stable, reliable electricity market'. www.alberta.ca/release.cfm?xID=44880BD97DCDC-D465-4922-25225F9F43B302C9 (last accessed June 2019).

——. 2016b. 'Alberta announces coal transition action', 24 November. www.alberta.ca/release.cfm?xID=44889F421601C-0FF7-A694-74BB243C058EE588 (last accessed April 2019).

——. 2016c. *Advisory Panel on Coal Communities.* www.alberta.ca/coal-communities.aspx (last accessed April 2019).

——. 2015a. 'Climate Leadership Plan'. Speech delivered by Premier Rachel Notley. 22 November. www.alberta.ca/release.cfm?xID=38886E9269850-A787-1C1E-A5C90ACF52A4DAE4 (last accessed April 2019).

——. 2015b. 'Climate Leadership Plan will protect Albertans' health, environment and economy', 22 November. www.alberta.ca/release.cfm?xID=38885E74F7B63-A62D-D1D2-E7BCF6A98D616C09 (last accessed April 2019).

Government of Canada. 2018. *Regulations Limiting Carbon Dioxide Emissions from Natural Gas-Fired Generation of Electricity.* http://gazette.gc.ca/rp-pr/p1/2018/2018-02-17/html/reg4-eng.html (last accessed April 2019).

Graney, E. 2018. 'Quick laws and freezing wages: Jason Kenney outlines plans for power', *Edmonton Journal*, 15 October. https://edmontonjournal.com/news/local-news/quick-laws-and-freezing-wages-jason-kenney-outlines-plans-for-power (last accessed April 2019).

Hussey, I & E Jackson. 2019. *Alberta's Coal Phase-Out: A Just Transition?* Parkland Institute.

International Energy Agency Statistics. 2014. *Electricity Production from Coal Sources (% of Total)*. The World Bank. https://data.worldbank.org/indicator/EG.ELC.COAL.ZS?year_high_desc=true (last accessed April 2019).

Kent, G. 2016. 'Coal industry fighting Alberta plans to phase out power plants', *Edmonton Journal*, 31 March. http://edmontonjournal.com/business/energy/coal-industry-fighting-alberta-power-plans (last accessed April 2019).

Landberg, R & A Hirtenstein. 2018. 'Coal is being squeezed out of power by cheap renewables', *Bloomberg*, 21 June. www.bloomberg.com/news/articles/2018-06-19/coal-is-being-squeezed-out-of-power-industry-by-cheap-renewables (last accessed April 2019).

MacArthur, E, N Poole, L Easton, M Fraser, L Hildebrand & P Leung. 2016. *Workers' Climate Plan: A Blueprint for Sustainable Jobs and Energy*. Iron & Earth and Energy Futures Lab. www.workersclimateplan.ca (last accessed April 2019).

Maclean's. 2017. 'Justin Trudeau's speech in Houston: Read a full transcript', 10 March. www.macleans.ca/economy/justin-trudeaus-speech-in-houston-read-a-full-transcript (last accessed April 2019).

Marr-Laing, T & B Thibault. 2015. *Early Coal Phase-out Does Not Require Compensation*. Pembina Institute. www.pembina.org/pub/early-coal-phase-out-does-not-require-compensation (last accessed April 2019).

Mertins-Kirkwood, H. 2017. *Evaluating Government Plans and Actions to Reduce GHG Emissions in Canada: Just Transition Policies*. Adapting Canadian Work and Workplaces to Respond to Climate Change & Canadian Centre for Policy Alternatives.www.adaptingcanadianwork.ca/canadas-evolving-domestic-climate-policy-landscape (last accessed April 2019).

Morgan, G. 2017. 'Alberta could be coal-free years ahead of deadline as ATCO plans transition to natural gas by 2020', *Financial Post*, 10 May. http://business.financialpost.com/commodities/energy/alberta-could-be-coal-free-years-ahead-of-deadline-as-atco-plans-transition-to-natural-gas-by-2020 (last accessed April 2019).

Natural Resources Canada. 2018. *Coal Facts*. Government of Canada. www.nrcan.gc.ca/energy/facts/coal/20071 (last accessed April 2019).

Ontario Power Generation. (n.d.). *Atikokan Biomass Conversion*. Ontario Power Generation. www.opg.com/generating-power/thermal/stations/atikokan-station/Pages/atikokan-station-biomass-conversion-project.aspx (last accessed April 2019).

Smith, R & K McDougal. 2017. *Costs of Pollution in Canada: Measuring the Impacts on Families, Businesses and Governments*. International Institute for Sustainable Development. www.iisd.org/library/cost-pollution-canada (last accessed April 2019).

Statistics Canada. 2018a. 'Table 25-10-0017-01: Electric power generation, annual fuel consumed by electric utility thermal plants'. Government of Canada. www150.statcan.gc.ca/t1/tbl1/en/tv.action?pid=2510001701 (last accessed April 2019).

——. 2018b. 'Table 25-10-0019-01: Electricity from fuels, annual generation by electric utility thermal plants'. Government of Canada. www150.statcan.gc.ca/t1/tbl1/en/tv.action?pid=2510001901 (last accessed April 2019).

——. 2018c. 'Table 36-10-0434-03: Gross domestic product (GDP) at basic prices, by industry, annual average (x 1,000,000)'. Government of Canada. https://ww150.statcan.gc.ca/t1/tbl1/en/tv.action?pid=3610043403 (last accessed April 2019).

——. 2018d. 'Table 36-10-0402-01: Gross domestic product (GDP) at basic prices, by industry, provinces and territories (x 1,000,000)'. Government of Canada. www150.statcan.gc.ca/t1/tbl1/en/tv.action?pid=3610040201 (last accessed April 2019).

——. 2018e. Table 36-10-0489-01: Labour statistics consistent with the System of National Accounts (SNA), by job category and industry. Government of Canada. www150.statcan.gc.ca/t1/tbl1/en/tv.action?pid=3610048901 (last accessed April 2019).

——. 2018f. 'Table 25-10-0046-01: Coal, monthly production and exports (x 1,000)'. Government of Canada. www150.statcan.gc.ca/t1/tbl1/en/tv.action?pid=2510004601 (last accessed April 2019).

——. 2018g. 'Table 14-10-0129-01: Union status by geography'. Government of Canada. www150.statcan.gc.ca/t1/tbl1/en/tv.action?pid=1410012901 (last accessed April 2019).

——. 2018h. 'Table 25-10-0014-01: Crude oil and equivalent, monthly supply and disposition (x 1,000)'. Government of Canada. www150.statcan.gc.ca/t1/tbl1/en/tv.action?pid=2510001401 (last accessed April 2019).

——. 2018i. 'Table 25-10-0063-01: Supply and disposition of crude oil and equivalent'. Government of Canada. www150.statcan.gc.ca/t1/tbl1/en/tv.action?pid=2510006301 (last accessed April 2019).

——. 2018j. 'Table 25-10-0047-01: Natural gas, monthly supply and disposition (x 1,000,000)'. Government of Canada. www150.statcan.gc.ca/t1/tbl1/en/tv.action?pid=2510004701 (last accessed April 2019).

——. 2017. 'Table 98-400-X2016290: Industry – North American Industry Classification System (NAICS) 2012 (427A), Class of Worker (5A), Labour Force Status (3), Age (13A) and Sex (3) for the labour force aged 15 years and over in private households of Canada, Provinces and Territories, census metropolitan areas and census agglomerations, 2016 Census – 25% sample data'. Government of Canada. https://tinyurl.com/yyjexxxf (last accessed April 2019).

The Canadian Press. 2016. 'Justin Trudeau: "There is no hiding from climate change"', *Maclean's*, 3 October. https://www.macleans.ca/politics/ottawa/justin-trudeau-there-is-no-hiding-from-climate-change (last accessed April 2019).

Thibault, B & A Read. 2016. 'Fact checking the coal industry's "information meetings"', *Pembina Institute Blog*, 2 March. www.pembina.org/blog/fact-checking-coal-industry-s-information-meetings (last accessed April 2019).

TransAlta. (n.d.). *Coal to Gas Conversions Project*. TransAlta. www.transalta.com/about-us/coal-to-gas (last accessed April 2019).

Varcoe, C. 2017. 'As province offers support for coal workers, employees wonder what future holds'. *Edmonton Journal*, 14 November. https://edmontonjournal.com/business/energy/varcoe-as-province-offers-support-for-coal-workers-employees-wonder-what-future-holds/wcm/fcd5e514-03db-4af8-a6d8-f416ac9a52e4 (last accessed April 2019).

Vriens, L. 2018. *The End of Coal: Alberta's Coal Phase-Out*. International Institute for Sustainable Development. www.iisd.org/library/end-coal-albertas-coal-phase-out (last accessed April 2019).

Wallace, R. 2001. *The British Columbia Advantage: Lessons from Alberta on the De-regulation of the Electricity Industry*. Parkland Institute. www.parklandinstitute.ca/the_british_columbia_advantage (last accessed April 2019).

Western Economic Diversification Canada. 2018. 'Minister Sohi announces funding to support Alberta communities transitioning to a low-carbon economy', *Cision News*, 16 November. www.newswire.ca/news-releases/minister-sohi-announces-funding-to-support-alberta-communities-transitioning-to-a-low-carbon-economy-700697481.html (last accessed April 2019).

Wilt, J. 2017. 'Six handy facts about Alberta's coal phase-out', *The Narwhal*, 17 January. https://thenarwhal.ca/six-handy-facts-about-alberta-s-coal-phase-out (last accessed April 2019).

9

Just transition solutions and challenges in a neoliberal and carbon-intensive economy

Darryn Snell

INTRODUCTION

While often forgotten, just transition's origins reside with the union movement and their commitment to social justice and fairness. Unions articulated a new vision for managing environmental protection that moved beyond the pitfalls of the traditional 'jobs vs the environment' dichotomy, of which unions so often became victim, by acknowledging the climate crisis and the need to urgently address it in a way that does not unfairly burden or disadvantage workers and local communities. In this regard, just transition is 'not just another transition' (ILO, 2018:1). At the core of union demands for a just transition is the desire for workers and communities to be protected from hardship during the urgent and desirable transition towards sustainable economies and societies.

As illustrated in other chapters in this book, the just transition concept has been appropriated, amended and repackaged by a range of other actors including environmentalists, business interests, governments, international organisations, academics and the popular media. This has resulted in considerable dilution in its original conceptualisation by trade unions (see Snell, 2018; Felli, 2014). While academic interest in just transition has tended to focus on documenting and categorising 'varieties' of labour environmentalism and just transition (e.g. liberal environmentalist, social reformist, grass-roots systems change) (see Goods, in this volume; Stevis et al., 2018), unions have been much more pragmatic and applied in their orientation.

Not surprisingly, unions have not expressed a lot of interest in the academic debates surrounding just transition but have sporadically questioned the orientation of the academic discourse and some of the underpinning assumptions such as the utility in categorising jobs as 'green' or 'dirty' (see Maher, 2016). Generally, however, unions have been focused on practical approaches and solutions for realising a just transition for their members and how best to achieve just outcomes through community and workplace activism, collective bargaining, and formal political engagement at the local, national and international levels (see CLC, 2000; Hampton, 2018; IndustriALL, 2018; ITUC, 2017; ITUC et al., 2006; Snell & Fairbrother, 2010; Stevis, 2018; TUC, 2008).

Union statements, media releases and reports on just transition have grown substantially in recent years. Policy positions are common in these documents with their demands being fairly clear and consistent between unions in different countries. Frequent union just transition demands include:

- Social dialogue and union involvement in the transition process;
- Embracing a commitment to comprehensive industrial policy making that ensures a sustainable transition including protecting as many existing jobs as possible and generating new ones;
- The establishment of government authorities to manage transition and economic revitalisation, including dedicated funds to upgrade facilities in impacted communities and economic diversification;
- Investment in new industries and technologies including 'green' energy which may also include 'clean' coal technologies;
- Benefits and assistance for displaced workers including wage support, enhanced retirement and redundancy packages, relocation assistance, and retraining, skills upgrading and employment support;
- Priority employment for displaced workers to take up decent jobs in other industries; and
- Social plans to protect the interests of disadvantaged workers and their families including mental health and well-being assistance.

Many of these demands depend upon a particular role and commitment from the state which is more akin to social democratic states of Europe, where industrial planning and social protection is enshrined and social partnership involving unions is the norm in many countries. In liberal market economies, there is no institutional guarantee that governments will uphold such commitments nor are they likely to place legal obligations upon employers to uphold socially responsible approaches to restructuring and facility closure as is common in many European countries (see Dekocker et al., 2011; European Commission, 2008; Forde et al., 2009; Stuart et al., 2007). How workers and communities achieve 'just transition' in liberal market contexts, therefore, presents its own particular challenges. This chapter focuses on this question through an examination of recent just transition initiatives pursued by unions in Australia's liberal market context.

On a per capita basis, Australia is one of the most carbon-intensive economies in the world and is under considerable pressure to make adjustments to its energy production mix (Diesendorf, 2009; Pearse, 2009). Australian unions took an early lead in making 'climate change union business' (ACTU, 2011; AMWU, 2008; CFMEU, 2011; ETU, 2009) but have often struggled to find meaningful and practical just transition solutions with its energy sector dominated by multinational corporations and governments committed to neoliberal policy making. This chapter documents these challenges for Australia's unions and discusses recent efforts to find fair and equitable solutions for workers impacted by the closure of one of the country's most polluting power stations. It draws upon interviews conducted with union leaders directly involved in just transition initiatives at the local level and their reflections on their perceived successes and ongoing challenges 18 months after the closure. I argue that realising a just transition for disadvantaged workers and communities in liberal market contexts presents particular challenges requiring ongoing union struggle, innovation and involvement post closure. These union activities, however, have the prospect of revitalising union purpose and their roles within communities.

CLIMATE CHANGE ACTION PARALYSIS AND AUSTRALIA'S MARKET-BASED ENERGY TRANSITION

In 2015, Australia joined more than 190 other nations in signing the Paris Climate Agreement (UNFCCC, 2015) and committed to curbing harmful carbon emissions 26–28 per cent by 2030 compared with 2005 levels. Australia's inclusion in the Paris Agreement was important for several key reasons: Australia is one of the highest per capita carbon emitters among western nations, and the world's largest exporter of coal forecast to be worth $AUS 58.1 billion in 2018–19 (Caruana, 2018). Indeed, Australia's vast fossil fuel resources have often been considered as foundational to Australia's economic success (Daniels, 1992; Gal & Brueck, 2018; Ville & Wicken, 2013). The international climate change crisis, matched with local concerns about the increasing incidents of droughts, bushfires, rising temperatures, coral bleaching along the Great Barrier Reef and rising sea levels, has left many commentators, environmentalists and concerned citizens questioning whether Australia's rich coal and natural gas resources should be considered the country's economic 'luck', 'dumb luck', or a 'curse' on future generations (see Cleary, 2011; Pearse, 2009). With economic growth and trade heavily hinged on natural resource extraction, Australia has struggled to decarbonise and introduce sustainable development proposals resulting in higher volumes of resource exploitation and environmental degradation (Daniels, 1992; McNeil, 2009).

As noted by Goods (in this volume, p. 88), 'climate change is a deeply toxic issue in Australian politics and society.' The result is ongoing political paralysis in relation to greenhouse gas (GHG) emissions reduction at the national level (Curran, 2012) despite a growing body of evidence suggesting Australia will fail to meet its Paris Agreement carbon emissions reduction targets (Skarbek, 2018). Whether Australia meets its international obligations under the Paris Agreement depends largely on government policy and the behaviour of private investors in GHG-emitting sectors.

The largest single source of GHG emissions is the electricity sector which is dominated by coal-fired power generation production. It is in this sector that we have seen the highest levels of political policy contestation and political division. Since 2000, various federal and state governments have introduced policies aimed at diversifying the energy supply, introducing more renewable energy production and setting GHG reduction targets. Most of these policy initiatives have been pursued by Australian Labor Party (ALP)-led Governments during which the Australian Green Party wielded some influence in getting legislation through Parliament and/or as a potential threat to certain key ALP electorates.

In 2007, the Kevin Rudd-led ALP Government was the first federal government that attempted to introduce a policy specifically aimed at tackling the nation's carbon emissions. Known as the Carbon Pollution Reduction Scheme (CPRS), the policy was designed as a cap-and-trade mechanism which introduced a price on carbon emissions and set an upper limit on the country's carbon emissions. However, despite repeated efforts to get the CPRS legislation through Parliament, the Rudd Government ultimately failed and ended up shelving the legislation in April 2010. In 2011, the Gillard-led ALP Government revised the CPRS proposal and succeeded in passing a Clean Energy Bill which introduced a carbon pricing and reduction policy framework including a programme aimed at the early retirement of some of the nation's most polluting power generation assets (Australian Government, 2011).

In the 2013 federal election, the Conservative Tony Abbott-led Coalition Government came to power and worked quickly to deliver on its election promise to repeal the Gillard Government's carbon reduction scheme. Under the Abbott Government, and the more recent Turnbull and Morrison Coalition Governments, fossil fuels have been given a reprieve with the government committing to provide additional support for 'clean coal' technologies and openly criticising some states for their 'reckless' pursuit of renewable energy targets (Scott & Williams, 2016). Scott Morrison, the current prime minister, even went so far as to bring a lump of coal into the Australian Parliament while serving as the Turnbull Government's treasurer to demonstrate the government's unwavering commitment to supporting the coal industry, including coal-fired electricity generation (for further discussion, see Goods, in this volume).

Despite attempts to prop up coal-fired generation by Coalition Governments, Australia's stationary electricity sector is in the midst of a significant transition to cleaner electricity sources. A number of state governments continue to support renewable energy targets and finance alternative energy projects which contribute to significant renewable energy uptake (Climate Council, 2017). Despite the uncertain political environment at the federal level, private investors continue to support renewable energy projects over fossil fuel projects. In addition, Australian consumers are busy installing photovoltaic electricity systems on their homes at some of the fastest rates in the world. As the Latrobe Valley case study highlights, owners of coal-fired power stations also continue to make decisions to divest away from carbon-polluting assets, making it difficult for the political backers of coal to get much traction in the changing market. Between 2012 and 2017, ten coal-fired power stations were decommissioned in Australia with no new coal-fired stations built or seriously considered (Senate Committee, 2017). As a result, a disconnect developed between government energy policy and the energy policy being pursued by powerful market actors such as private owners and financiers of stationary power stations (see Snell & Schmidt, 2012).

Most recently, the Morrison Coalition Government has sought to turn the tide by promising to finance and build new coal-fired power stations resulting in strong rebuke from the stationary energy sector. The Australian Energy Council, the peak employer body for the electricity and natural gas industry, warned that the government-supported investment ran the risk of undermining investment in new facilities or reinvestment in existing plants (Preiss & Carey, 2018). While Australia's business leaders are far from united in their views and approach to addressing climate change (see Goods, in this volume) a significant number of companies across all sectors continue to call upon government to bring about stability to the energy and climate change policy debate, with many of Australia's largest industrial and mining firms calling for a price on carbon as a way to bring stability to the market. So despite Prime Minister Morrison's calls for more coal-fired power stations and better support for coal exports, Australia's low-carbon transition, as in most countries, refuses to retreat. Much of this is due to renewables becoming much more competitive as an energy source in recent years as a consequence of rapid technological change and private and public investment being redirected towards alternative energy.

This context of energy policy uncertainty, political division and paralysis surrounding emissions reductions – in a sector dominated by large, privately owned electricity generation businesses driven by market principles – presents unique challenges for realising a just transition for those workers and communities disrupted by Australia's ongoing energy transition. Trade unions in Australia's Latrobe Valley power generation region have come to know these challenges all too well in recent years as ageing power stations are shuttered and workers displaced. The following discussion presents the views and insights of Latrobe Valley union leaders who led the struggle for a just transition in this market-based context.

THE HAZELWOOD CLOSURE AND JUST TRANSITION IN THE LATROBE VALLEY

The Hazelwood Power Station closure in March 2017 presented a major opportunity for the adoption of just transition approaches for managing Australia's shift to a low-carbon economy. Responsible for the employment of some 750 workers, including those working for contractors, the Hazelwood Power Station closure would be the largest decommissioning of an Australian coal-fired power station to date. The history of the Hazelwood Power Station tells the story of Australia's decisive move towards liberal market policy position since the 1980s (Weller & O'Neil, 2014). The 1,600-megawatt station was built by the Victoria Government-owned State Electricity of Victoria (SEC) between 1964 and 1971 in the Latrobe Valley. The region is located about a hundred miles east of Melbourne and contains vast reserves of lignite (or 'brown coal' as it is locally called) that were first exploited soon after the First World War to provide inexpensive electricity for the state's industrialisation.

During its lifetime, the SEC operated as a vertically integrated public utility, owning and operating open-cut coal mines and associated power stations in the Latrobe Valley and power distribution and retailing businesses throughout the state (Galligan, 1998). Up until April 2017, the Latrobe Valley's three vast open-cut coal mines (Yallourn, Loy Yang and Morwell) and four brown coal-fired power stations (Hazelwood Power Station, Yallourn Power Station, Loy Yang A Power Station, and Loy Yang B Power Station) were responsible for generating over 85–90 per cent of the state's electricity. However, the privatisation of the SEC and associated Latrobe Valley mines and power generators, including Hazelwood, in the

1990s meant that the industry came to be almost exclusively owned by multinational energy corporations headquartered outside Australia (see Beder, 2009; Snell & Schmitt, 2012). According to the Latrobe Valley's power industry unions, it was the privatisation of the industry which presented the most significant challenges for managing the transition to a low-carbon economy and assisting disadvantaged Hazelwood workers.

> ... if it was a SEC, in other words a government owned power industry, it [just transition] would be a no-brainer ... if we hadn't have been privatised we probably would've had two more stations operating on what we call gasification of the coal which would've been a lot more environmentally friendly ... what they [Liberal Government] did was flog off the power stations and the new owners they've done minimum maintenance and run them into the ground ... they don't give two shits about people in this region or the community. (GTLC secretary)

Prior to its closure, the Hazelwood Power Station had been a target for a range of environmental organisations who labelled the station the 'dirtiest' power plant in the Southern Hemisphere (Baer, 2010; Ker, 2010; Wilson, 2010). On various occasions, the 'Close Hazelwood' campaigns resulted in tensions between the local community, who viewed Hazelwood as an important employer and supporter of community organisations, and environmentalists who locals largely considered 'outsiders' due to their headquarters and activist base being located in Melbourne and other large cities. Like any region, the Latrobe Valley had its share of committed environmentalists and environmentally concerned citizens, but most stayed clear of coal-related issues due to the local sensitivities and associated employment opportunities fossil fuels provided. Most local union leaders also viewed the campaign as that of 'outsiders' who did not understand the region's economic dependence on coal nor the power station's importance in maintaining inexpensive baseload power for the country. While conversations were held between local union and environmental organisations, including about the closure of Hazelwood, they kept a safe geographic and political distance from one another, and refrained from expressing open hostility or appreciation for each other's positions.

In 2011, environmentalists became increasingly optimistic that their hope of securing the closure of the Hazelwood Station would be realised as the Gillard-led ALP Government began to pursue a policy aimed at bringing about the early retirement of some of the nation's most polluting power generation assets (Australian Government, 2011). Commonly referred to as the 'contract for closure' programme, three

Latrobe Valley power plants, including Hazelwood, entered into negotiations with the Gillard Government about the purchase and closure of
their assets (*Latrobe Valley Express*, 2009). The purchase price requested
by the private owners of the power generators, however, was deemed
as unreasonable by government negotiators and the programme ended
with minimal outcomes much to the disappointment of environmentalists (McRae, 2012a,b). Latrobe Valley residents, workers and unions,
however, expected Hazelwood's reprieve would be short lived given the
age of the station, lack of investment by its multinational owners and
ongoing environmental concerns about its carbon emissions.

Ultimately, however, Hazelwood's demise would come about not as
a consequence of environmental activism or government policies but
due to the corporate business decisions made by its overseas owners. In
November 2016, the Paris-based energy company Engie and its investment partners (Mitsui and Co. of Japan) announced their decision to
close their jointly owned Hazelwood Power Station. According to Engie's
public statements, the company made this decision in order to reduce its
carbon emissions and involvement in coal-fired power generation and
as a consequence of the power station's age, high operating costs and
the substantial investment needed to upgrade the facility. The announcement caught the state government and local Latrobe Valley community
by surprise. As one Latrobe Valley union organiser noted, 'I think Daniel
Andrews [Premier of the State of Victoria] and the Labor Government
at the time were caught a little bit off-guard by what Engie did. They
knew it was coming, but I don't think they were ready for the earliness of
when it was called' (ETU organiser). But, according to a different union
representative. they had been

> … tipped off by one of the company managers' about the decision
> weeks before the announcement. Even so the unions expected Engie
> to adopt a staged closure of Hazelwood's eight generation units over
> several years rather than the abrupt full closure with less than five
> months to prepare for it. Unions strongly condemned this approach
> to closure and singled out this company decision as the greatest
> impediment to realising a just transition for Hazelwood workers;
> 'Six months' notice with Hazelwood was probably the single biggest
> hurdle. (CFMEU M&E assistant secretary)

Power industry unions and the region's Labour Council, the Gippsland
Trades and Labour Council (GTLC) were not unfamiliar with the challenges ahead, as they had lived through the privatisation of the power
industry in the 1990s when thousands of jobs were lost and regional

unemployment became one of the highest in the country. As a consequence of this history, the region's unions began to prepare for a closure
scenario in the early 2000s through the holding of climate change and
just transition fora, discussions with government ministers and depart-
ments about closure and just transition programmes, and the careful
monitoring of investment and maintenance being carried out by the
power industry's corporate owners (Snell, 2018). The years of work discussing and preparing for industrial change in the power sector by local
unions and the local community ahead of the Hazelwood announcement
meant unions, the GTLC and the state government were fairly well positioned to develop a rapid response. The following section discusses the
major components of this just transition rapid response and how unions
perceive this response some 18 months after the closure.

THE WORKER TRANSFER SCHEME

A 'Worker Transfer Scheme' (WTS) developed by the three main power
industry unions (Construction, Forestry, Mining and Energy Union
(CFMEU), the Australian Manufacturing Workers Union (AMWU),
and the Electrical Trades Union (ETU)) and involving, the Australian
Services Union and the Association of Professional Engineers, Scientists
and Managers (Professional Australia) was a key component of Latrobe
Valley union-led just transition strategies. According to the unions
involved, the WTS represented the first attempt at introducing such a
scheme in Australia. The aim of the WTS was to open up job opportunities and facilitate transfer of ex-Hazelwood workers into employment at
the remaining Latrobe Valley power stations through an early voluntary
retirement scheme. Getting the scheme off the ground was not without
its challenges. At the time of the Hazelwood closure, unions were in a
protracted industrial dispute with the region's largest power station, Loy
Yang A, and commencing bargaining at other stations. In addition, the
other power stations were not particularly interested in the proposal and
the prospects of losing their skilled retiring workers. The workforces of
these other power stations also had reservations about how the scheme
might jeopardise their career progression and were initially resistant to
its introduction:

> The worker transfer scheme, I will say to you, has created some, how
> would you describe it, disgruntledness between other people and
> people from Hazelwood and when I say that is that some people have
> a perception that the worker transfer scheme is all about, 'Yeah, the
> Hazelwood people are being looked after and then they've got the

golden handshake already, being looked after, guaranteed a job at one of the other power companies and again, paid … .' (ETU organiser)

Eventually these diverse barriers were overcome through union negotiations with concerned members, union media pressure on the power stations to support the scheme, the involvement of a former Federal Minister serving as a facilitator between unions and owners of the remaining power stations and the involvement of the Victoria Labour Government who agreed to provide financial support to the remaining generators that redeployed Hazelwood workers.

On 10 March 2017, all power station owners, unions and the government signed the Latrobe Valley Worker Transfer Partnership Agreement whereby the 'parties acknowledged the significant job losses consequent upon the closure of the Hazelwood Power Station and mine', but also an 'opportunity to retain and refresh the Latrobe Valley power sector workforce to ensure its sustainability for the future' (*Latrobe Valley Worker Transfer Partnership Agreement*). Under the Agreement, the participation of social partners was completely voluntary and power station owners reserved their right to employer discretion in the selection and hiring of workers. In circumstances where candidates were considered as having equal merit, the companies agreed to preference ex-Hazelwood workers for the relevant position. However, power station owners could only access government support for the early retirement packages if they recruited ex-Hazelwood workers. According to union representatives, employers received $AUS 75,000 in government support for each Hazelwood worker they employed with the government allocating some $AUS 20 million to the scheme. Based upon their estimates, unions hoped the scheme would open up job opportunities for around 250 ex-Hazelwood workers; roughly half the Hazelwood workforce. Unfortunately, this was not to be the case.

By May 2018, only 65 ex-Hazelwood workers had benefited from the WTS with some 65 per cent transferring to Loy Yang A. While the scheme was intended to assist both the direct employees of Hazelwood as well as those working for Hazelwood contractors, unfortunately ex-Hazelwood contract workers tended not to benefit from the scheme; 85 per cent of beneficiaries of the WTS had been direct employees of Hazelwood. The average age of transferring workers was 42 years. This relatively older profile meant workers who were too young to retire but likely to confront significant age discrimination and retraining barriers could hopefully see out their remaining careers within the power industry performing similar work with comparable remuneration.

Unions blame the power station owners for the relatively poor outcomes from the scheme. In the case of Loy Yang, they maintain that the power station used the early retirement scheme as a 'job cutting tool' (CFMEU M&E assistant secretary) to downsize rather than re-deploy ex-Hazelwood workers. In addition, they claim that when power stations filled positions opened up by early retirement offers, they did not always employ ex-Hazelwood workers. In this regard, the remaining power stations have used the early retirement scheme to downsize, while also renewing their workforce with a mix of ex-Hazelwood workers and other recruits. In addition, unions also maintain that some of their ex-Hazelwood members have confronted discrimination by other power stations under the WTS due to their past union activities. The general feeling among unions involved is that the 'scheme's sort of been scuttled by the generators' (CFMEU M&E assistant secretary). This also caused many ex-Hazelwood workers to feel let down. According to one union representative, this resentment was partially due to the unreasonable expectations created by the scheme which were poorly managed and workers failing to understand 'there were no guarantees' (ETU organiser). However, for those ex-Hazelwood workers that did secure power station work under the scheme, it was a good outcome, both for them and the company in its bid to revitalise its ageing workforce. One ETU organiser involved in negotiating the WTS with the generators noted: 'I know a number of people who got work under the worker transfer scheme and they couldn't be more grateful for what those packages did.'

ASSISTING DISPLACED WORKERS

A central plank of union-based just transition proposals is support for displaced workers in re-skilling and/or finding alternative employment. Ideally, this support should be provided prior to closure and as part of a larger outplacement support programme post closure. According to union representatives, Hazelwood provided minimal support to employees prior to closure and the short closure notice meant workers struggled to achieve much benefit out of this assistance. In response to this situation, the GTLC worked to build capacity to provide outplacement assistance. The Labour Council's office was centrally located in the town of Morwell not far from the Hazelwood Power Station. In November 2016, the Council made the decision to open its office and staff it with volunteers who could provide support and guidance to Hazelwood workers on where they could access employment assistance. By the time of closure in March 2017, three hundred Hazelwood workers had sought support from the Council. Maintaining such support on a volunteer basis in the longer term, however, was not viable and the GTLC's secretary opened up conversations with the state government and the newly established Latrobe Valley Authority (LVA) about financial support for the provision of outplacement support for Hazelwood workers and their families (see Victoria Government, 2017). By March 2017, an agreement had been reached between the GTLC and LVA to form a partnership supporting soon-to-be retrenched Hazelwood workers under the banner of the WTS with public financial support provided to the GTLC for the delivery of this service. Involving local employment service and training providers, the WTS was made available to all employees of Hazelwood Power Station, associated contractors, supply chain employees and their family members. In May 2017, moreover, the scope of the WTS was broadened to include some 160 workers impacted by the closure of a local timber mill as well thus extending the labour market intermediary work of the Labour Council.

The WTS was set up in the GTLC and LVA offices to provide one-on-one transition services and advice related to skills, training, financial advice and employment assistance. The Labour Council's service was based on a peer-to-peer support model and ex-Hazelwood workers were employed as peer support officers. According to the assistant secretary of the GTLC, the selection of the peer-to-peer support worker was critical:

> ... our project team that's been involved at the Trades and Labour Council from day one have all been people that have had multiple employment. Those guys have worked on construction jobs that finished, worked in maintenance that have finished, worked as casual shut downs, whatever else. And understand the network you need, the contacts you need, the effort you need to make and the steps you need to take to get employment.

These peer support officers worked to identify the needs, challenges and interests of workers and advocate for them by negotiating with training providers to offer training courses and introducing them to other employment service providers. It was expected that many Hazelwood workers would struggle to find alternative ongoing employment due to their age (a large proportion of Hazelwood's workers were 40–50 years of age), many not having formal qualifications and highly specialised skills easily transferable to other industries as well as little work experience outside of Hazelwood (Snell et al., 2015). The WTS provided by the Labour Council aimed to engage these workers and their family members early on. By October 2018, some 1,398 workers and family members had registered and received assistance through the WTS. In an effort to improve its capacity to deliver this support, the Labour Council partnered with a non-profit community organisation specialising in providing employment assistance to job seekers in areas related to résumé writing, job search and interviewing techniques. Co-located in the GTLC office, this partnership worked to provide a 'one-stop' shop for displaced workers looking for assistance. Through further negotiations with the state government, the Labour Council also managed to secure dedicated retraining funds for displaced workers wanting to acquire new skills.

Adopting a case management model, GTLC peer support workers evaluated the existing skills of workers, the types of jobs they were interested in pursuing and the actual employment opportunities for these jobs, and areas where training support might be beneficial. The Labour Council then negotiated with reputable training providers to deliver relevant training courses for workers. Between April 2017 and July 2018, some 1,270 workers received specialised induction training brokered through the GTLC in areas such as construction, oil and gas, electrical, rail, industrial decommissioning and asbestos removal. Through the completion of induction training, these workers improved their chances of gaining entry into these emerging job areas. Other workers secured work in a diverse range of occupations in transportation and logistics, agriculture and healthcare. While there are no precise numbers on job outcomes, government figures estimate some 55 per cent of those seeking work after six months of closure had found employment, albeit often in part-time or short-term positions (Minister for Jobs, Innovation and Trade, 2017).

Representatives of the GTLC and affiliated unions were also careful to point out that the support services they provided also included looking after the mental health and well-being of workers as they transitioned to other employment opportunities: 'the social connection was important. And we still see it now where we're nearly two years down the track that people, when they're at a loose end, and we've encouraged it from day one, to drop into the Gippsland Trades and Labour Council, have a talk to the project officer there, have a cuppa and just have a chat' (GTLC assistant secretary).

JOB STIMULUS AND REGIONAL REVITALISATION

Regional revitalisation and job creation in impacted communities is an integral part of just transition but success is often difficult to achieve in the short to medium term. It is also an area where unions have limited capacity to deliver this outcome alone, although their involvement in regional revitalisation initiatives is critical. In the case of the Latrobe Valley, unions have advocated for and been closely involved in regional revitalisation and job stimulus discussions since the Hazelwood closure. In the short term, unions maintained that jobs for some displaced Hazelwood workers could be created through mine rehabilitation and power station decommissioning work. Engie agreed to this arrangement and employed some 135 Hazelwood workers to conduct work in these two areas. Unions expected the work to last for two years but most workers were made redundant after twelve months due to the early completion of the power station decommissioning and the decision by Engie to contract out remaining mine rehabilitation work.

Since the closure of the Hazelwood Power Station, the Victoria Labour Government has made longer-term regional diversification and revitalisation a priority, with the LVA being tasked with much of this responsibility. A major focus of these government-led economic revitalisation efforts has been to advance 'green' jobs although it is not explicitly expressed in these terms. One of the initiatives, for example, has been to establish a New Energy Jobs Fund aimed at stimulating jobs and investment in the energy sector, particularly in relation to renewables. Support for the development of a locally designed and engineered water turbine and a waste-to-energy project using anaerobic digestion are examples of the projects funded under the scheme. In addition, the Victoria Government supported the establishment of a workers' cooperative manufacturing solar hot water systems in the Latrobe Valley and negotiated with a Melbourne-based electric vehicle company to establish a Latrobe Valley-based factory estimated to create five hundred manufacturing jobs in coming years (Carey & Preiss, 2018). These initiatives fit well with just transition notions that there must be a shift towards 'green' jobs (Snell, 2018). In response, the union supported WTS is now assessing the jobs and skill requirements needed to support the firms involved in these initiatives, so ex-Hazelwood workers can be retrained to take advantage of these emerging job opportunities. However, it is important to stress that the region continues to have some of the highest unemployment in the state and the majority of jobs that are being created are in healthcare and community services where there are limited opportunities for ex-Hazelwood workers.

Regional revitalisation is going to be a long-term project particularly when it is expected that the ageing Yallourn Power Station will close in the next five to ten years. The LVA is increasingly inspired by European economic development approaches based on 'smart specialisation' (see Foray, 2014) and social partnership between industry, unions, government and other stakeholders (see Reitzenstein et al., this volume). This has ensured that unions, represented by the GTLC, take part in key economic development steering committees involved in the development of regional and industrial strategy. Unions are generally positive about these initiatives but there is a perception and concern that a considerable amount of money appears to be going to employ consultants to develop 'regional revitalisation' plans with questionable job benefits. The GTLC secretary who represents unions on a number of these regional development committees states:

> ... the only people who ever made any money out of that, and I actually talk about them a fair bit, are the consultants which I sort of see as snake oil salesmen and that run away and do a feasibility study and come up with their graphs and their charts and all their bullshit, but at the end of the day they come back and say, guess what? The business case don't stack up.

It is also perceived that there has been minimal engagement by local employers in these government-funded initiatives, which is problematic in a market-based economy where the private sector is expected to drive new employment opportunities. As illustrated by employer behaviour in relation to the WTS, expecting employers to actively support the aims of just transition and deliver relevant outcomes, even when subsidised by government, is often not enough. Overcoming these structural issues associated with employer control over just transition outcomes in liberal market economies is a major challenge. Success requires a much higher level of intervention on the part of the government and must be supported by a stronger just transition policy framework which forces private firms to better support local communities in transition through local hiring, redeployment and retraining practices.

DISCUSSION AND CONCLUSION

> One of the biggest problems we have in this region is that there are some little pressure groups around here, as I call them, who are pretty much the glass half empty all the time, you know the world is going to end. That's not our position. Our position as a Trades and Labour Council and the unions in this region is that we see a lot of opportunity going forward and we're going to try and maximise that opportunity as we'll talk to whoever we have to and follow whatever lead we can to try and make sure that we get jobs in this region; that people are confident and we're going ahead. (Secretary of the GTLC)

The GTLC secretary makes a significant, albeit often overlooked, point about the foundation of just transition needing to be built on one of optimism despite the evidence of major social, community and economic disruption. Also alluded to in the Labour Council secretary's reflection is that this optimism must be intertwined with a commitment to ongoing struggle aimed at achieving just outcomes for those disadvantaged.

Latrobe Valley power industry unions are proud of the just transition initiatives that they were able to secure for ex-Hazelwood workers. In particular, they are pleased to have developed and secured the first worker transfer scheme in Australia's history. However, they express significant disappointment and frustration with the lack of serious commitment by the region's private-sector actors in delivering better outcomes for ex-Hazelwood workers. This disappointment is expressed in relation to the behaviour of not only the remaining power stations but also firms in other industries. The unions had hoped to extend the WTS to other large non-power industry employers in the region, including the local pulp and paper mill, that employed workers with similar skill sets (e.g. welders, boilermakers, fitters and turners, electricians, etc.) but they were met with strong resistance from company owners. In a liberal market context where governments are hesitant to interfere in company decision making, these are the challenges for realising a just transition.

As noted by Weller and O'Neill (2014), not all liberal market economies are alike nor are the views and positions of governments who preside over them (see Mertins-Kirkwood & Hussey, in this volume). The current Andrews Victoria Labour Government has taken a keen interest in working with unions and the community in assisting disadvantaged Hazelwood workers and the revitalisation of the Latrobe Valley region. The Andrews Government is also well respected by the union movement for its approach to involving trade unions as a social partner. In the Latrobe Valley, this is demonstrated in its support for the WTS and the important work of the GTLC. In recognition of the positive work carried out by the GTLC assisting disadvantaged workers in the region, the Andrew Government has extended its support to the GTLC until the end of 2019 and continues to involve unions in local regional and industry planning. The government's keen interest in European-inspired approaches to just transition and regional development has made it much easier to realise many of the common union demands associated with just transition.

In the November 2018 State election, the Andrews Labour Government was returned to office with an increased majority. This victory should guarantee a level of continuity in state government commitment to the region and ongoing union involvement in the process. Unions like to point out that had a conservative coalition government been in power the situation would have been dramatically different, with little protection from market forces and the impacts of private sector decisions. 'You know to be perfectly honest', an AMWU organiser remarked, 'the Liberal Party [Australia's major conservative political party] model is based on market forces; stand or fall on your own. Now there's no way this community would be in the position it's in now if we'd had a Liberal Government in power.'

Achieving just transition in liberal market economies, where private-sector interests control valuable power station infrastructure, presents significant challenges when the practice of just transition depends upon an interventionist state committed to social partnership and union involvement in the transition process (see Rietzenstein et al., in this volume). For many of Australia's Latrobe Valley unions, the lessons to emerge from attempts to implement just transition for Hazelwood workers is that plant closure decisions cannot be left to the market. As unions speculate on when the next power station closure is likely to occur, they highlight the biggest problem in making this prediction is that closure decisions will most likely be made overseas without local consultation: 'Yallourn Power Station is owned by China Light and Power. The day that somebody sitting in a boardroom over there [Hong Kong] decides that it's not making enough profit, they can make that decision overnight' (GTLC secretary).

In light of the Hazelwood closure and the failure of its private owners to provide early notice and a staged closure, and the other power stations to fully commit to the WTS they signed up to, unions are working to get improved legislation over future closures. They are calling for the establishment of a federal government independent authority to manage future power station closures through helping to 'co-ordinate and stagger power station closures and support the transition of workers – whether to other power stations slated for later closure, retraining for other industries or early retirement' (CFMEU, 2018:1). The current conservative Federal Coalition Government is unlikely to back such a proposal and while the opposition Labor Party has expressed verbal support for such legislation, there is no guarantee that they will if they take control of Government. The liberal market context will continue to require unions to find innovative ways to achieve just transition through the bargaining process, involvement of local communities and engagement with power station owners and government. Their proactive initiatives, activism and engagement in just transition solutions at the local level has not only renewed their purpose but also provided important lessons for unions confronting similar challenges elsewhere. Latrobe Valley union leaders are not climate sceptics nor are they denying the fact that the world is currently moving – albeit at a slow speed – away from fossil fuels. They do however maintain that more needs to be done to achieve just transition and the challenges confronting workers in carbon-exposed industries are only going to increase in years ahead. What is clear to unions is that the next power station closure is not far away and that despite Prime Minister Morrison's calls for more coal-fired power stations to be built and better support for coal exports, Australia's low-carbon transition, like most countries, is inevitable. Unions therefore need to prepare for an increasingly carbon-constrained world. The Latrobe Valley unions' approach to tackling these challenges, however, is very pragmatic. Local union leaders have little interest in debates about what does and does not qualify as a just transition:

> I'm over these acronyms and just transition and green jobs and all that stuff, let's just get down to the basics. The basics are this, one industry's shut down, workers have lost their jobs, they're looking at any other industry where they can get a job. Now depending on what the make-up of that industry is, I mean you know if people say, well unless you get a green job you failed, I mean who are they kidding? I mean for a worker, a job is a job, right? There's going to be opportunities out there. The industry's going to change, there's no doubt about that ... Because we understand jobs mean prosperity, it means future for our kids and going forward (GTLC secretary).

This focus on 'going forward' and preparing the next generation for new types of jobs has renewed the union movement's sense of purpose and drives much of the union movement's ongoing efforts for just transition and sense of optimism in this rapidly changing regional context.

REFERENCES

ACTU (Australian Council of Trade Unions). 2011. *Climate Change is Union Business*. Melbourne: ACTU.

AMWU (Australian Manufacturing Workers Union). 2008. *Making Our Future: Just Transitions for Climate Change Mitigation*. Granville, Australia: AMWU.

Australian Government. 2011. *Securing a Clean Energy Future*. Canberra: Commonwealth of Australia.

Baer, H. 2010. '"Switch off Hazelwood" Rally', *Green Left Weekly*, no. 857, 15 October. www.greenleft.org.au/content/%C3%A2%C2%80%C2%98switch-hazelwood%C3%A2%C2%80%C2%99-rally (last accessed April 2019).

Beder, S. 2009. 'Electricity, generation, climate change and privatisation', *Australian Options*, vol. 57, pp. 18–20.

Carey, A & B Preiss. 2018. 'New factory to make Morwell "electric car capital" as Labor eyes marginal seat', *The Age*, 30 October. www.theage.com.au/politics/victoria/new-factory-to-make-morwell-electric-car-capital-as-labor-eyes-marginal-seat-20181030-p50ctn.html (last accessed April 2019).

Caruana, L. 2018. 'Coal overtakes iron ore as Australia's largest export', *Australia's Mining Monthly*, 3 July. www.miningmonthly.com/investment/international-coal-news/1341648/coal-overtakes-iron-ore-as-australia%E2%80%99s-largest-export (last accessed April 2019).

CFMEU (Construction, Forestry, Mining and Energy Union). 2018. 'New independent authority needed to manage transition for energy workers'. https://me.cfmeu.org.au/news/new-independent-authority-needed-manage-transition-energy-workers (last accessed April 2019).

——. 2011. *Why a Price of Carbon is Essential for our Future*. CFMEU.

CLC (Canadian Labour Congress). 2000. *Just Transition for Workers During Environmental Change*. Ottawa: Canadian Labour Congress. https://digital.library.yorku.ca/yul-1121737/just-transition-workers-during-environmental-change/datastream/OBJ/download (last accessed April 2019).

Cleary, P. 2011. *Too Much Luck: The Mining Boom and Australia's Future*. Collingwood, Victoria: Black Inc.

Climate Council. 2017. *Renewables Ready: States Leading the Charge*. Canberra: Climate Council of Australia.

Curran, G. 2012. 'Contested energy futures: shaping renewable energy narratives in Australia', *Global Environmental Change*, vol. 22, no. 1, pp. 236–244.

Daniels, P. 1992. 'Barriers to sustainable development in natural resource-based economies: Australia as a case study', *Society and Natural Resources*, vol. 5, no. 3, pp. 247–262.

Dekocker, V, V Pulignano & A Martens. 2011. '"Acting out" institutions: A cross-sector analysis of local unions' response and practices of resistance to collective redundancy in Belgium', *Employee Relations*, vol. 33, no. 6, pp. 592–606.

Diesendorf, M. 2009. *Climate Action: A Campaign Manual for Greenhouse Solutions*. Sydney: UNSW Press.

ETU (Electrical Trades Union). 2009. *Job Creation: The Case for a National Gross Feed-in Tariff*. Melbourne: ETU Victoria Branch,

European Commission. 2008. *Restructuring in Europe 2008 – A Review of EU Action to Anticipate and Manage Employment Change*. Brussels: European Commission.

Felli, R. 2014. 'An alternative socio-ecological strategy? International trade unions' engagement with climate change', *Review of International Political Economy*, vol. 21, no. 2, pp. 372–398.

Foray, D. 2014. *Smart Specialisation: Opportunities and Challenges for Regional Innovation Policy*. London: Taylor and Francis.

Forde, C, M Stuart, J Gardiner, I Greenwood & R Mackenzie. 2009. *Socially Responsible Restructuring in an Era of Mass Redundancy*. Working Paper 5. Leeds: Centre for Employment Relations Innovation and Change, Leeds University Business School.

Gal, S & H Brueck. 2018. 'Canada is richer than the US, according to a new wealth ranking – in fact, the US doesn't even make the top 10', *Business Insider*, 7 November. www.businessinsider.com/richest-countries-in-the-world-2018-10?r=US&IR=T (last accessed April 2019).

Galligan, B (ed.). 1998. *Local Government Reform in Victoria*. Melbourne: State Library of Victoria.

Hampton, P. 2018. 'Trade unions and climate politics: Prisoners of neoliberalism or swords of climate justice?' *Globalizations*, vol. 15, no. 4, pp. 470–486.

ILO (International Labour Organization). 2018. *Just Transition Towards Environmentally Sustainable Economies and Societies for all*. Geneva: ILO.

IndustriALL. 2018. 'Spanish coal unions win landmark Just Transition deal'. www.industriall-union.org/spanish-coal-unions-win-landmark-just-transition-deal (last accessed April 2019).

ITUC (International Trade Union Confederation). 2017. *Just Transition: A Report for the OECD*. www.oecd.org/environment/cc/g20-climate/collapsecontents/Just-Transition-Centre-report-just-transition.pdf (last accessed April 2019).

——, ETUC (European Trade Union Confederation) & TUAC (Trade Union Advisory Committee to the OECD). 2006. *Trade Union Climate Change Strategies: The Trade Union Statement to COP12/MOP2*. Nairobi: UNFCCC.

Ker, P. 2010. 'Hazelwood protest has support of key greens', *Sydney Morning Herald*, 12 September. www.smh.com.au/environment/hazelwood-protest-has-support-of-key-greens-20090911-fkv4.html (last accessed April 2019).

Latrobe Valley Express. 2009. 'Pay up to power down', 19 October, p. 1.

Maher, T. 2016. 'Why Australia needs a Just Transition', *CFMEU Mining and Energy Division*. https://me.cfmeu.org.au/leadership-message/why-australia-needs-just-transition (last accessed April 2019).

McNeil, B. 2009. *The Clean Industrial Revolution*. Sydney: Allen and Unwin.

McRae, L. 2012a. 'More fears for workers', *Latrobe Valley Express*, 3 May, p. 2.

——. 2012b. 'Call to end "contracts" charade', *Latrobe Valley Express*, 30 August, p. 1.

Minister for Jobs, Innovation and Trade. 2017. 'Former Hazelwood workers get a new start', 30 August. www.premier.vic.gov.au/former-hazelwood-workers-get-a-new-start/ (last assessed April 2019).

Pearse, G. 2009. 'Quarry Vision: Coal, Climate Change and the End of the Resources Boom', *Quarterly Essay*, vol. 33, pp. 1–122.

Preiss, B & A Carey. 2018. 'Energy firms cool on Guy's pledge to build new power plant'. *The Age*, 13 November, p. 8.

Scott, J & P Williams. 2016. 'Turnbull steps up attack on renewables after Australian blackout', *Bloomberg News*, 29 September.

Senate Committee. 2017. *Retirement of Coal-fired Power Station: Final Report*. Canberra: Environment and Communications Reference Committee, Commonwealth of Australia.

Skarbek, A. 2018. 'Australia is not on track to reach 2030 Paris target (but the potential is there)', *The Conversation*, 6 September.

Snell, D. 2018. '"Just transition?" Conceptual challenges meet stark reality in a "transitioning" coal region in Australia', *Globalizations*, vol. 15, no. 4, pp. 550–564.

—— & P Fairbrother. 2010. 'Unions as environmental actors', *Transfer*, vol. 16, no. 4, pp. 411–424.

—— & D Schmitt. 2012. '"It's not easy being green": electricity corporations and the transition to a low carbon economy', *Competition and Change*, vol. 16, no. 1, pp. 1–19.

——, D Schmitt, A Glavas & L Bamberry. 2015. 'Worker stress and the prospects of job loss in a fragmented organisation', *Qualitative Research in Organisations and Management*, vol. 10, no. 1, pp. 61–81.

Stevis, D. 2018. 'US labour unions and green transitions: Depth, breadth, and worker agency', *Globalizations*, vol. 15, no.4, pp. 454–469.

——, D Uzzell & N Räthzel. 2018. 'The labour-nature relationship: Varieties of labour environmentalism', *Globalizations*, vol. 15, no. 4, pp. 439–453.

Stuart, M, C Forde, R MacKenzie & E Wallis. 2007. *An Impact Study on Relocation, Restructuring and the Viability of the European Globalisation Adjustment Fund: The Impact on Employment, Working Conditions and Regional Development*. Brussels: Policy Department, Economic and Scientific Policy, European Parliament. www.europarl.europa.eu/thinktank/en/document. html?reference=IPOL-EMPL_ET(2006)385647 (last accessed April 2019).

TUC (Trades Union Congress). 2008. 'Trade unions and climate change: A just transition', TUC Climate Change Conference, 16 June. London: Congress House.

UNFCCC (United Nations Framework Convention on Climate Change). 2015. *Adoption of the Paris Agreement. Twenty-first session of the Conference of the Parties*. UN Doc. No. FCCC/CP/2015/L.9/Rev.1. 12 December.

Victoria Government. 2017. 'Landmark schemes gives Hazelwood workers a fresh start', 10 March. www.premier.vic.gov.au/landmark-scheme-gives-hazelwood-workers-a-fresh-start/ (last accessed April 2019).

Ville, S & O Wicken. 2013. 'The dynamics of resource-based economic development: Evidence from Australia and Norway', *Industrial and Corporate Change*, vol. 22, no. 5, pp. 1341–1371.

Weller, SA & P O'Neil. 2014. 'An argument with neoliberalism: Australia's place in a global imaginary', *Dialogues in Human Geography*, vol. 4, no. 2, pp. 105–130.

Wilson, J. 2010. 'Tight security at Hazelwood protest', *ABC Gippsland News*, 10 October. ww.abc.net.au/local/photos/2010/10/10/3034179.htm (last accessed April 2019).

Notes on contributors

Kali Akuno is the Co-Founder and Executive Director of Cooperation Jackson, an emerging network of worker cooperatives and supporting institutions in Jackson, Mississippi. Kali is an organiser, educator, writer for human rights and social justice and a co-editor of *Jackson Rising: the Struggle for Economic Democracy and Black Self-Determination in Jackson, MS* (Daraja Press, 2017). Kali has authored numerous articles and pamphlets including 'The Jackson-Kush Plan: the Struggle for Black Self-Determination and Economic Democracy'.

Martín Álvarez Mullally is currently based in Roca (Río Negro, Argentina) as a researcher at the Observatorio Petrolero Sur (OPSur). He coordinates the Río Negro *Mesa de Transicion Post Petrolera* and is the author of *Alto Valle Perforado* (Ed. Del Jinete Insomne, 2015) a book that looks at the social implications of expanding unconventional fossil fuel exploitation. He works with different social organisations in northern Patagonia.

Fernando Cabrera Christiansen is a Social Communicator by training. His research and work in Argentina focuses on the socio-territorial transformations caused by the intensification of fossil fuel exploitation. He coordinates the Observatorio Petrolero Sur (OPSur) and the Enlace por la Justicia Energética y Socioambiental (Socioenvironmental and Energy Justice Alliance, EJES).

Caleb Goods is a Lecturer of Management and Employment Relations at the University of Western Australia Business School. His research focuses on the challenges of interlinking the greening of work and society. Caleb is also the chief investigator of a research project examining work and workers experiences in the 'gig economy'.

Felix Heilmann is a Research Assistant on climate and energy policy at E3G and a student at the University of Oxford's Department of Politics and International Relations. His work concentrates on the net zero transition in Germany and Europe, with a special focus on designing and delivering just transition pathways.

Ian Hussey is a Research Manager at the University of Alberta's Parkland Institute and a Steering Committee Member of the Corporate Mapping Project (a Partnership Grant funded by the Social Sciences and Humanities Research Council of Canada). Ian is the co-author of two forthcoming reports on Alberta's phase-out of coal-fired electricity.

Dunja Krause is a Research Officer at the United Nations Research Institute for Social Development (UNRISD), where she leads the work on social dimensions of climate change. A geographer by training, she coordinates the Institute's work on just transition and conducts research on transformative adaptation in coastal cities.

Laura Maffei is Co-coordinator of the Chico Mendes Socioenvironmental Department of the Union of Education Workers of Río Negro (UnTER), Argentina. A civil engineer by training, she teaches at the National University of Río Negro and, in collaboration with social organisations and trade unions in Latin America, monitors different environment-related issues.

Hadrian Mertins-Kirkwood is a senior researcher with the Canadian Centre for Policy Alternatives where he focuses on international trade and climate policy. He is a member of several just transition research and advocacy networks and is a public commentator on just transition issues in the Canadian context.

Edouard Morena is a Lecturer in French and European Politics at the University of London Institute in Paris (ULIP). Co-coordinator of the Just Transition Research Collaborative (JTRC), his current research concentrates on the social and justice dimensions of climate action through a focus on non-state actors' involvement in the international climate debate. He is the author of *The Price of Climate Action: Philanthropic Foundations in the International Climate Debate* (Palgrave, 2016) and co-editor, with Stefan Aykut and Jean Foyer, of *Globalising the Climate: COP21 and the Climatisation of Global Debates* (Routledge, 2017).

Nils Moussu is part-time lecturer in International Political Economy and associate member of the Centre of International History and Political Studies of Globalization at the University of Lausanne. His research focuses on business associations' political strategies in climate change governance, and business narratives on the environment.

Alexander Reitzenstein is a policy advisor at E3G Berlin. He leads on E3G's work on just transition and coordinates work on climate legislation, coal and sustainable finance in Germany and Central Eastern Europe. Alexander has earned Master's degrees in public policy from the London School of Economics and Political Science and Hertie School of Governance.

Anabella Rosemberg is the International Programme Director at Greenpeace International. From 2004–17, she was the Environment Policy Officer for the International Trade Union Confederation (ITUC). In that position, she coordinated the international trade union movement's policy making on climate and other environmental issues and led trade union advocacy efforts in various intergovernmental processes (UNFCCC, Rio+20), securing, among others, the inclusion of just transition in the Paris Agreement on climate change.

Sabrina Schulz is a policy expert with a background in climate, energy and industrial policy as well as sustainable finance. She currently heads the Berlin office of KfW, the German national promotional and bilateral development bank. Before, she worked as Director of Third Generation Environmentalism (E3G) in Berlin and in various policy capacities in Germany, the UK, the US and Canada. She earned her PhD in International Politics from Aberystwyth University in the United Kingdom.

Darryn Snell is an Associate Professor in the School of Management at RMIT University in Melbourne, Australia. Co-coordinator of the Skills, Training and Industry Research Group, he has developed a stream of research focused on labour and economic transitions in carbon-exposed regions. Through his applied research and trade union activities, he works closely with unions and governments on finding practical 'just transition' solutions for workers disadvantaged by environmental policies and industrial restructuring.

Dimitris Stevis is Professor of Politics at Colorado State University. His research focuses on global labour and environmental politics, with particular attention to labour environmentalism and social and ecological justice. He is a founder of the Environmental Justice Working Group within the School of Global Environmental Sustainability. He has recently co-edited, with Nora Räthzel and David Uzzell, a special issue of the journal *Globalizations* (15:4) entitled 'Labour in the Web of Life'.

Sandra van Niekerk is pursuing a PhD on energy transitions at a local government level at the University of the Western Cape. She also works on the energy democracy programme at the Alternative Information Development Centre, South Africa, and supports research on public services in Africa for the Public Services International (PSI) global trade union federation. Previously, Sandra worked for many years as the National Education Officer for the South African Municipal Workers' Union.

Index